# Taildragger Tactics

## By Sparky J. Imeson

Aurora Publications
SAN: 297-5262 ISBN: 1-880568
P.O. Box 573
Jackson, Wyoming 83001-0573 USA

# Taildragger Tactics

**By Sparky J. Imeson**

Published by:

Aurora Publications
SAN: 297-5262 ISBN: 1-880568
P.O. Box 573
Jackson, Wyoming 83001-0573 USA


First Edition
First printing: April 2004
Printed in the United States of America
ISBN 1-880568-70-5 $24.95 Softcover

*This book is dedicated, with love:*

*To my mom:*

**Jennie F. Imeson**

*To the memory of my dad:*

**Paul C. Imeson**

# Acknowledgments

For the past 35 years, scores of people have offered their advice, input, thoughts and assistance in attempting to teach the author how to master the conventional-gear airplane.

The late Howard Ballew, Jackson, Wyo. in teaching me mountain flying techniques and taildragger operations, suffered through more bounced landings in his Cessna 180, with the patience of Job, than any friend should have to endure.

The late Dave Low, Pinedale, Wyo. spent hours discussing the aerodynamics, techniques and procedures for dealing with taildragger operations.

I was not alone in creating *Taildragger Tactics*. Many other people throughout the years have provided ideas, techniques, practice and guidance. A special thank you goes to my father, the late Paul Imeson, for his direction, assistance, patience and encouragement.

After reading the transcript, a multitude of suggestions were made by Brian Jones, Parker, Colorado, Reno Brown and Rusty Brown, Jackson, Wyoming, Jeannie MacPherson and Bill Gallea, Helena, Montana. To each I express my humble thanks.

And finally, gratitude goes to my mother, Jennie Imeson, who as front-line editor carried the burden of correcting the readability and grammatical errors.

Cover photos courtesy of:

Stuart Horn<br>
President<br>
Aviat Aircraft Inc.<br>
PO Box 1240<br>
Afton, WY 83110

# About the Author

Sparky Imeson was born and raised in Jackson, Wyo. His grandfather homesteaded at the confluence of the Hoback and Snake Rivers south of Jackson.

The 'flying bug' snuck up and bit him and he began flying at Jackson in 1966, in his father's Piper J-3 Cub and Cessna 205.

While attending the University of Wyoming, Laramie, as a predental candidate, Sparky worked at the Chem-Zoo library, a service station and pizza house for the first two years. He then became a police dispatcher, advanced to patrolman and patrolman specialist. He was accepted at the University of Oregon dental school. His father, Paul, flew from Jackson to Laramie to ferry Sparky home for short visits before moving to Oregon. Sparky decided he liked flying better than looking in someone's mouth. He attended Herrod Aviation, Billings, Mont. to become a professional pilot.

In 1969, as part-owner of Imeson Aviation at the Jackson Hole Airport, he became involved with the AOPA Wyoming fly-in that traveled to Jackson, providing a handout on mountain flying recommendations.

Sparky became alarmed at the number of aircraft accidents attributed to mountain flying. When he couldn't find a book on the subject, he sat down and wrote *Mountain Flying*.

Sparky has had a diverse education in aviation, being directly involved in airplane sales and rentals, fueling, maintenance, flight training, on-demand charter, air ambulance, agricultural aerial application, mosquito spraying, aerobatic training, demonstration pilot, and corporate flying.

Sparky received the FAA Northwest Region Flight Instructor of the Year Award in 1974, 1979 and 1995, for providing effective and creative flight and ground instruction. He spent one year as editor of *Wings West* magazine, before learning that the job of editor affords very little time for flying. Of his 19,200+ flight hours, the majority has been in small airplanes in the mountains—which is his first love after his daughters.

## WARNING AND DISCLAIMER

Our society is constantly changing and perhaps it is not for the better. While growing up in Jackson, Wyo. a handshake, rather than a written contract, was the only requirement for a 'deal.' Homes were left unlocked. Keys were left in auto ignitions. No one ever messed with another's property.

That's all changed. Now our society requires a warning-disclaimer for books that provide suggestions of a "how-to" nature. This is due to the legal profession. Don't peg me with a prejudicial attitude toward lawyers. I worked as a law enforcement officer for 5 1/2 years and as a pilot-paralegal for another 10 years. Some of my best friends are attorneys.

This book is an attempt to collect the information necessary to operate the taildragger airplane safely under various conditions of flight. As a result it is a subjective view, colored by my own experiences, prejudices and requirements. Much of my flight experience has been in mountain flight operations at high-elevation airports. Since my requirements for taildragger flying may differ from your circumstance, some techniques that have worked well for the author may not meet your needs or experience level. Without background, education, training, experience and practice, some of the procedures and techniques can be very dangerous.

Pilots with a desire to learn advanced techniques, such as the *curved-path takeoff* or *curved-path landing*, should do so only with the help of a knowledgeable and qualified taildragger instructor.

The purpose of this book is to provide thought-provoking information in the format of a general guide for ground and flight operations. The book is not intended to replace a flight instructor in learning taildragger operations. All information is presented "as is," without warranty of any kind, either express or implied, including but not limited to the implied warranties of suitability or fitness for a particular purpose.

The user assumes the entire risk as to the accuracy and the usefulness of this book. Mountain Flying LLC, Aurora Publications and the author shall have no liability nor responsibility to any person or entity for the loss or damage that is caused or alleged to be caused, directly or indirectly, by the information, maneuvers or procedures contained within this book.

If you do not agree with the above and do not wish to be bound by the above, you may return this book to the publisher or author for a full refund.

# Taildragger Tactics
## TABLE OF CONTENTS

# 10 COMMANDMENTS

1. **Always remain in a position where you can turn to lowering terrain** when approaching or flying over mountains.
2. When flying upslope terrain, **Do not fly beyond the point of no return,** defined as the position where, if the throttle is reduced to idle, the aircraft can be turned around during a glide without impacting the rising terrain.
3. If **71 percent of the takeoff speed is obtained at the halfway point** of a runway, the airplane will takeoff in the space remaining.
4. **Never enter a canyon if there is not room to turn around.**
5. **Always approach for landing at the same indicated airspeed** regardless of altitude.
6. Study weather trends and conditions before takeoff.
7. **Approach ridges and mountains at a 45-degree angle** to provide an escape route.
8. **Do not thermal shock or detune** the engine.
9. Prepare and carry in an accessible area an **emergency survival kit.**
10. **Avoid becoming complacent.** Do not ignore warning signs of weather, terrain or wind.

# Introduction

## INTRODUCTION

A book about flying should be simple; it should be void of complicated mathematical formulas. A book about taildraggers is no exception. The book should be readily understandable. Whenever a complicated mathematical formula is required for an explanation, that formula should be reduced to a usable form or rule of thumb. Formulas requiring a calculus background or hours of intensive study in order to determine what that formula is trying to prove may be scientifically accurate, but it is preferable to translate most formulas to experience and common sense. This avoids educational gaps and misunderstanding that can result because the reader skips certain information as a result of the complexities of mathematics. Formulas, when presented in this book, reduce mathematics to just a few key pieces of information.

## STRUCTURE

The person purchasing a taildragger book is interested in the 'meat' of flying the conventional-gear airplane. My initial reaction was to structure this book in the same manner as a computer-program application book. That is, to do away with all the introductory information and list the steps concerned with how to install the program, how to start the program, and then how to use it. Supplemental information could be provided for more detailed operations, review and delving into the intricacies of the program. Nevertheless, the initial objective would be to get the program up and running.

After trying this approach it was determined that the elimination of the introductory information is not practical when dealing with a taildragger. Certain flight techniques are predicated upon previous knowledge. Without imparting this information to insure that all pilots have been exposed to it, flight could be dangerous. Playing with a computer before learning the proper procedures has none of the potential consequences of playing with a taildragger when you are not properly prepared.

This book is structured to allow you to read from beginning to end, like a novel. Most of the chapters build and expand upon information that was presented in the previous chapters. However, each chapter is also self-contained to allow you to read only those chapters desired.

## FLIGHT OPERATIONS

Flying a taildragger requires knowledge and skill. The knowledge is imparted in this book. The skill is something that must be developed in the airplane.

All flying involves risk. That is part of the glamour of flight, but reducing the risk does not reduce the glamour. This book may help you deal with situations beyond the limits of your experience or capabilities and in this way it reduces your risk. Some "tricks" are presented that may help when you encounter situations that go beyond the capabilities of the pilot or the airplane.

Aside from being fickle, Mother Nature can be a mean old lady. She can provide challenges to test the skill of the best pilots. With knowledge, practice and common sense, Mother Nature can, for the most part, be subdued.

What strikes me as the greatest trouble spot for pilots flying taildraggers (tricycle-gear airplanes too), is a lack of understanding of the aerodynamic principles involved during the slow flight regime and the characteristics the pilot can expect from the airplane while operating at the minimum controllable airspeed. This, too, can be learned. With the number of new and prospective taildragger pilots on the horizon, we have a lot of learning to do.

## AVIATION ACTIVITY

Back in 1970 when Mountain Flying was written, I remarked "Aviation activity has increased substantially during the past few years and barring some catastrophe, it will continue to increase throughout the coming years."

The number of pilots increased steadily from 1970 until the late 1980s, then began to decline. Some people feel the passage of the aviation "statute of repose" is going to revive general aviation. Although, psychologically, this legislation should help, the influx of $100,000-plus-trainer airplanes is not the answer.

---

**A man who doesn't know his limitations has never put himself to the test.**

---

Aviation activity is making a comeback. Rather than the number of new start pilots, aviation is returning as a resurgence in tail-dragger training. This rebirth began before the statute of repose. The increased taildragger activity may be due to pilots wanting to expand their experience, safety and competence. It may also be due to the cutback in manufacturing of small, affordable airplanes. This has caused a proliferation of home-built aircraft with tail wheel configurations. Many pilots are switching from airplanes with "training wheels" to conventional-gear airplanes. It is important to have a basic understanding of the aerodynamics involved with the operation of these taildragger airplanes.

## FLYING THE TAILDRAGGER

There are those pilots who say, "If you don't fly a taildragger, you ain't #*!%." This is a bunch of malarkey. Taildraggers are not that much harder to fly than tricycle-gear airplanes; they are different. Once the differences are mastered, the pilot has no more trouble operating the taildragger than the tricycle-gear plane, despite the fact that the taildragger may have stricter limitations concerning crosswind operations.

To fly a taildragger safely, one must be aware of and comply with two limitations. You must know and recognize your limitations as a pilot, and operate within the limitations of the airplane.

In this respect, finding a flight instructor who allows you to make mistakes is important for your training. The instructor must allow you to make errors and teach you how to correct those errors, because it is not your mistakes in aviation that will "get you," it's how you take care of them.

**Concerning taildraggers: Fly the darn thing until it's tied down.**

The taildragger pilot is often described as a safer pilot than the tricycle gear pilot. Frequently this is true, because more taildragger pilots learn to use the rudder, not only on the ground, but in the air. Proper control coordination enhances the safety of flight.

It has been said, "A man who doesn't know his limitations has never put himself to the test." This applies to taildragger operations. When a pilot begins his tail wheel training, the first few takeoffs may resemble the antics of a staggering drunk. Overcontrolling with rudder pressure and with the brakes will cause the airplane to swerve back and forth across the runway. This is normal for the beginning pilot because of the lack of recognition of the aerodynamic forces working against the airplane. Without experience, knowing when the rudder should be applied to correct for a swerve that has not yet begun, is mighty difficult. Soon, though, with practice, the pilot begins to anticipate the need for corrections and makes them before the need arises. The takeoff straightens out and the pilot relaxes ... but not too much. One thing can be said with certainty about taildraggers: "Fly the darn thing until it's tied down."

## AERODYNAMICS

It is my hope that the preceding has convinced you why a chapter about aerodynamic forces has been included; after all, I realize that all you want to do is read the chapter on landings and get on with the taildragger flying. But, with an understanding of the turning forces affecting the airplane on the ground, not only is the airplane easier to control, but during a "tight," one's knowledge may save the airplane.

The following chapters contain advice, warnings, data, rules of thumb, and possible courses of action for the pilot to keep in mind while operating the taildragger. The book has been divided into chapters that represent the phases of flight (preflight, taxiing, takeoff, en route, arrival and landing) to make it easy to find what you need when you need it.

---

**A tight is a situation that occurs, in which you have no prior experience, but your former knowledge, understanding and practice of similar maneuvers or circumstances can be transferred or applied to cope with this new situation.**

---

## NON-ESOTERIC STYLE

The author has been accused of providing too much material about subjects that are often considered basic knowledge. In an attempt to avoid an esoteric style and to help you understand what is happening and why it is happening, I plead guilty.

A full understanding of any particular maneuver or procedure is basic to the elimination of, "Do it this way, because I say so," method of learning to fly. A book that can be understood by only a chosen few is worthless in learning to fly. The purpose of this book is not just to arouse your curiosity, but to satisfy it. In doing so, I have found I am not writing so much to answer your questions, but rather my own, providing an explanation that satisfies myself.

## COMPLACENCY

In addition to answering your questions and satisfying my own, it is important to fly the airplane at all times, even on the ground.

---

**Complacency: a lack of judgment and overconfidence arising from familiarity and boredom in performing a particular task over and over.**

---

Webster's defines complacency as a "calm sense of well-being and security," or "self-satisfaction accompanied by unawareness of actual dangers or deficiencies." Smug, familiar, bored, overconfident, and self-contented are adjectives describing complacency. It has been determined that a lack of judgment and overconfidence often results when a task is repeated over and over.

One of the best ways to determine whether or not you are becoming complacent is to scrutinize the manner in which you perform your preflight walk around. Do you really examine the airplane, or are you paying less attention to detail? The same thing applies to the actual flight. Do you think ahead of the airplane, planning for all contingencies? If not, you might be setting yourself up for an accident. Remember: complacency kills!

## CONVENTIONS

Icons are used throughout this book to ascertain that certain information stands out and to aid in sorting out nice-to-know information from need-to-know information, areas of knowledge that should be remembered, along with advisory, precautionary and warning information.

*The "Advanced" icon has been placed to serve as a warning that the procedure or maneuver should not be attempted by inexperienced pilots without first seeking flight instruction from a competent instructor.*

*The "note" icon denotes information that you should know or, at least, spend a moment reading the content; however, it is not critical to the operation of the taildragger.*

*The "remember" icon marks information that you will use over and over. It should be remembered to enhance the safety of flight.*

*The "Rule of Thumb" icon points out procedures that save time and trouble. They are roughly correct and do not detract from safety, but they are not meant to be scientifically accurate.*

*The "warning" icon represents a situation, event or area that can get you in trouble (serious trouble) or cost you money (airplane damage), if you do not adhere to the recommendations.*

# CHAPTER 1

# Aerodynamic Forces

## INTRODUCTION

Flying a taildragger safely is predicated upon the pilot having a greater knowledge of aerodynamics. The additional knowledge is not overly complicated or drawn out, but it *is* required information.

> **The Cub is the safest airplane in the world, it can just barely kill you.** *—Max Stanley, Northrop test pilot*

The classic question, "Why do you climb a mountain?" is often followed by the response, "Because it's there." This question and answer may be applied equally to flying the taildragger.

Why do *you* want to fly a taildragger?

Mountain climbing is considered a challenging and dangerous activity. Before a climber can be considered experienced and qualified enough to go out on his own, he begins taking lessons. He must learn the basics ... without any shortcuts. He then progresses on his endeavor one-step-at-a-time, increasing his experience level with each outing.

Flying the taildragger is also a potentially dangerous activity. It should also be tackled one-step-at-a-time. For the pilot who has not been exposed to an understanding of the aerodynamics of flying, there is a chance that he, like the mountain climber, can stumble and fall. The seriousness of the fall depends on the regime of flight and the extent of his knowledge.

---

> **"This thing we call luck is merely professionalism and attention to detail, it's your awareness of everything that is going on around you, it's how well you know and understand your airplane and your own limitations. Luck is the sum total of your abilities as an aviator. If you think your luck is running low, you'd better get busy and make some more. Work harder. Pay more attention. Study your NATOPS[1] more. Do better preflights."** *—Stephen Coonts, The Intruders.*

---

This section is meant to ensure exposure to the essential building blocks of aerodynamic information. Pilots can be compared to buildings; if the foundation is faulty, the building will eventually fail.

Some of the important information includes stalls, spins and twisting forces. Information concerning airborne flight and ground operations is included here. The detailed aerodynamics of ground operations, such as taxiing, takeoff, landing and crosswind operations, is presented in the appropriate sections. Some of it is basic, but again, it is included to insure you have been exposed to this information.

### Intuition

The purpose of providing aerodynamic information can be described by one word, *"intuition."* This noun is described as the *"power or faculty of attaining to direct knowledge or cognition*

[1] NATOPS - *Naval Air Training and Operating Procedures Standardization*

*without evident rational thought and inference."* This "quick and ready insight" is gained from a knowledge of aerodynamics and may be useful in getting yourself out of a "tight."

When a pilot is receptive to additional knowledge, and has been provided with that information, he may, without conscious effort, be able to apply that wisdom to new and different situations.

This isn't the same thing as *instinct*, for you will find that instinct is often wrong when flying an airplane. Take for instance the spin. The spin is an aggravated stall that begins when yaw causes the outside wing to circumscribe a larger radius of turn than the inside wing, thus providing the lift for the rotational movement. Instinct in a spin demands that you pull back on the control wheel to recover. Pulling back has always worked before, but in this circumstance it aggravates and maintains the spin. A "conditioned reflex" must be used to break the stall and fly out of the resulting dive.

Next to weather accidents, stall/spin accidents cause the majority of general aviation accidents and the greatest number of serious injuries and deaths. Having an understanding of aerodynamics is, therefore, important in avoiding stall/spin accidents.

I used to maintain that stalls are not scary, especially when one knows what is going on and is able to predict what the airplane is going to do beforehand. I went out with one guy who was really afraid of stalls, and after a few stalls, I was too.

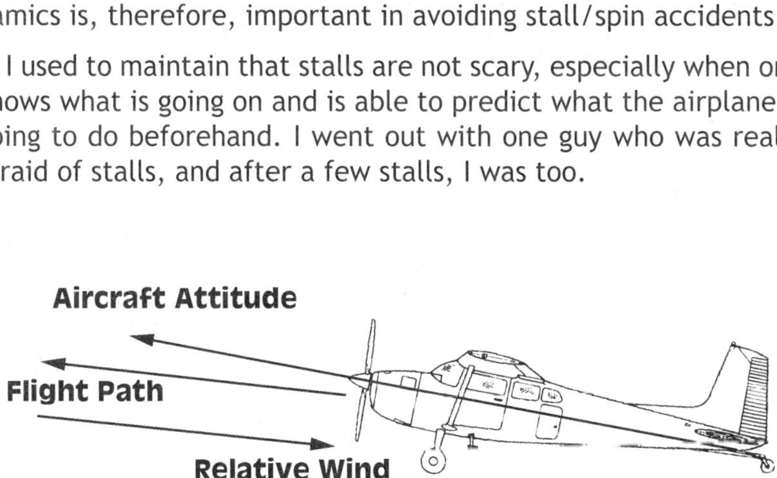


*Figure 1.* The relative wind is parallel and opposite from the flight path. General aviation airplanes are thrust limited. Because excess power is not available, the flight path may not parallel the attitude of the airplane. Although the nose of the airplane may be pointed up in the air (relative to the horizon), the actual flight path, or climb, will be less than the airplane's attitude.

Seriously, stalls used to bother me when I was learning to fly. I diligently practiced them, even though I did not want to. Solo practice (even dual instruction) of stalls caused me a great deal of consternation. The reason is because the darn airplane did something different each time it stalled. One day, at the moment of the stall, the wing would rotate and fall off to the right; the next day to the left. Occasionally, with power on, both wings would fall through the horizon in a level attitude, but not often enough to lend any level of comfort to the practice of stalls.

To relieve this fear, a basic knowledge of stall aerodynamics is necessary. It is very easy to understand why an airplane reacts the way it does during stalls ... thank goodness. Boy, what a relief.

---

> **The course of the flight up and down was exceedingly erratic, partly due to the irregularity of the air, and partly to lack of experience in handling this machine.** — *Orville Wright*

---

The typical general aviation pilot is taught that lift is created by a pressure differential explained by Bernoulli's Principle. That is, "When the velocity of a fluid is increased, its pressure is reduced." The fluid can be a liquid or a gas. Stressed to a lesser degree is Newton's Third Law of Motion that "every action has an equal and opposite reaction." This explains how the wing pushes down on the air and the equal and opposite reaction is for the air to push up on the wing. It is a combination of both that allows the airplane to fly.

It appears that nowadays some fledgling pilots have not been exposed to the effects of the four twisting forces on flight as they affect the airplane during a stall. This becomes most evident during stall practice and recovery. During a climb with insufficient rudder application, the ball of the turn-and-slip indicator is to the right of the index (assuming clockwise rotation of the propeller as viewed from the cockpit). What will the airplane do at the moment of the stall? Knowing this relieves apprehension.

So why hasn't this been taught before? It's not that old timers in aviation didn't know anything about the aerodynamics of flight. Most of them did. It may have something to do with the removal of the requirement for spin instruction and demonstration for the private pilot certificate. Wolfgang Langewiesche in *Stick and Rudder* (published in 1944), presented an excellent treatise on aerodynamics that should remain required reading for all pilots to this day.

Another possible area of blame is that some instructors like to take shortcuts to (they think) speed up the learning process. It's not these time savers in teaching that causes the omission of the important information. Rather, it's an *attitude problem* that has caused many important items of learning to fly to be omitted from our education.

Some aviation authors felt that knowing aerodynamics in depth was not necessary. Others felt it was beyond a pilot's capability to understand. And perhaps it is, if it is explained as a mathematical model. So we won't use mathematics. I promise ... really.

## SLOW FLIGHT AND STALLS

With an understanding of the aerodynamic forces involved in landing a taildragger, problems can be corrected without experiencing any dire consequences. This is especially important if the flare is initiated too high, the airspeed gets too low, or a wing drops before the landing.

**Man must rise above the Earth — to the top of the atmosphere and beyond — for only thus will he fully understand the world in which he lives.**
— *Socrates*

By comprehending the aerodynamics involved, the pilot will pick up a dropped wing with the rudder or coordinated use of aileron and rudder, rather than trying with the aileron alone (instinct again). The use of ailerons alone may lack sufficient aerodynamic force to rotate the wings to a level attitude before contact with the ground or they may even aggravate the situation at a slow speed.

Airplanes manufactured since the late 1940s are required by regulation to have some aileron control during a stall. This is accomplished by modifications to the wing. Some wings are "twisted," others incorporate stall strips.

If a straight wing is mounted to the fuselage at a predefined angle of incidence and then some mechanical contrivance grasps the wing tip and twists it toward the front of the airplane, the tip area would have a smaller angle of incidence than the inner portion of the wing. During flight, when the wing root area reaches the critical angle of attack, the outer portion of the wing is still flying.

Other manufacturers use a stall strip to retain aileron control during the stall. This is a small triangular-shaped strip attached to the leading edge of the wing somewhere between the wing root and the mid portion of the wing's length. When approaching the critical angle of attack, the strip disrupts the smooth flow of air and causes the inner portion of the wing to stall before the wing tip area can stall.

---

> **The air up there in the clouds is very pure and fine, bracing and delicious. And why shouldn't it be? — it is the same the angels breathe.** — *Mark Twain, 'Roughing It,' Chapter XXII, 1886*

---

The use of aileron alone to pick up a dropped wing is not desirable. For example, if the left wing drops, the natural reaction or instinct is to move the stick to the right. This is the proper aileron input, but it is ineffective. The left aileron is deflected down, increasing the curvature and lift of the wing. This lift, especially at the slow speed of landing, may not be adequate to create the lift to cause a roll. This is because lift and drag are directly proportional. As the lift increases, so does the drag. The increased drag resulting from the down aileron deflection may aggravate the situation, yawing the nose left and causing the wing to drop further.

At slow speeds the flight controls lose effectiveness in the order of A-E-R, that is, aileron, elevator, and then rudder. A pilot learning this will use the rudder to pick up a dropped wing close to the ground at slow airspeed since the aileron and elevator do not help. It is important to know this.

The use of ailerons is instinctive, and in this case it is the wrong action. The use of rudder must be a conditioned response, with training in the same way Pavlov's dogs were conditioned to salivate when hearing the dinner bell.

One example of conditioning is through spin training. When the airplane enters a spin you are convinced the nose is pointing straight down (it's probably somewhere just over halfway between the horizon and straight down). The normal reaction is to pull back on the stick to raise the nose. It's always worked before. But now the stick must be moved forward to break the stall, then eased back at a rate that will not exceed the critical angle of attack, to return to level flight.

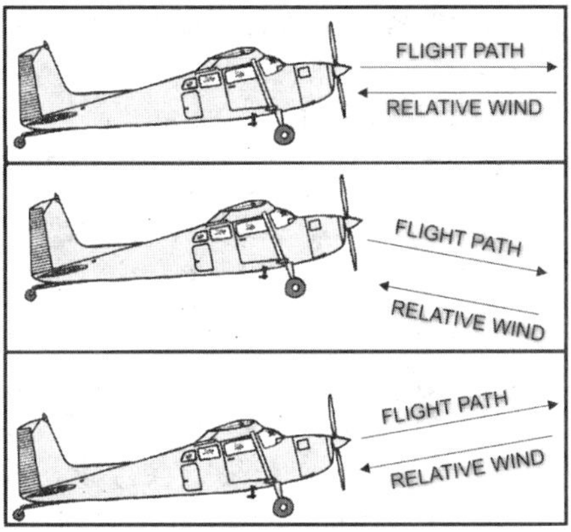


*Figure 2.* **The relative wind is parallel and opposite from the flight path.**

Some people are not comfortable with slow flight and stalls. In taildragger operations, it is important that the pilot is comfortable so he can direct his full attention to the operation of the airplane, and not have a nagging doubt in the back of his mind about

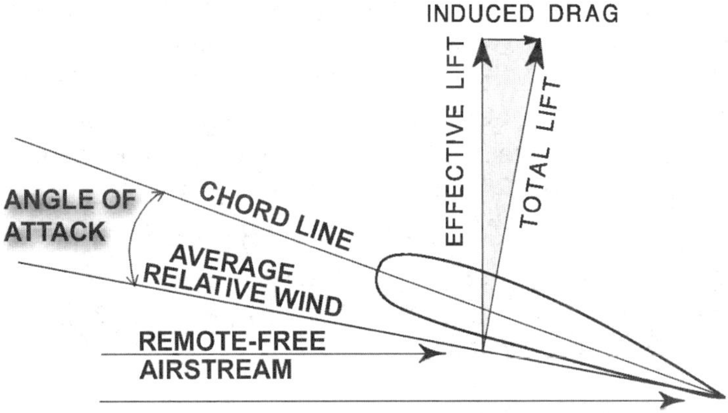


*Figure 3.* Angle of attack - the angle between the chord line of the wing (a line connecting the most forward part of the wing to the most rearward part of the wing) and the relative wind (created by movement of the wing through the air. It is parallel and opposite the flight path). Total lift acts perpendicular to the wing. The effective lift opposes the weight. The rearward component between the effective lift and total lift is the induced drag (associated with lift—the greater the lift, the greater the induced drag). Free airstream is the airstream unmodified by the wing and fuselage.

what the airplane is going to do or not do under certain situations. If you are not at ease with slow flight, flying a taildragger is like dancing with a porcupine. Even if you put your feet in the right places, it's hard to concentrate on the music.

The following is basic aerodynamic information. Although it is basic, consider, as a minimum, taking a look at the section on rudder usage during stalls.

---

> **Most gulls don't bother to learn more than the simplest facts of flight — how to get from shore to food and back again. For most gulls, it is not flying that matters, but eating. For this gull, though, it was not eating that mattered, but flight. More than anything else, Jonathan Livingston Seagull loved to fly.** — *Richard Bach, 'Jonathan Livingston Seagull'*

---

## BASIC AERODYNAMICS

Explaining aerodynamics requires some basic definitions. During the design of an airplane the engineers want the cockpit area to be level when in cruise flight. They do this by attaching the wing to the fuselage at an angle that will produce sufficient lift to sustain level flight at cruise power settings. This small angle at which the wing is attached to the fuselage is called the *angle of incidence*. It is fixed. The pilot cannot change it.

The *relative wind* is created by the movement of the airplane on the ground or in the air. It is "relative to" the airplane's ground path or flight path. An understanding of relative wind is basic to understanding aerodynamics.

## FLIGHT PATH

Take for example an airplane flying with a nose-up attitude of about 8-10 degrees relative to the horizon.

This airplane may be in level flight, called slow flight, with the nose-high attitude. A high-power setting is required to maintain level flight.

This same airplane may be operating with reduced power and it is descending while maintaining the same attitude as before.

Or, with excess power above that required for level flight, it is climbing.

In each instance the 'attitude' of the airplane remains the same but the relative wind changes, based upon the flight path.

The longitudinal axis may represent the chord line of the wing in the mind of some pilots. When the airplane is pointed up, the pilot expects the airplane to go up. But we've just seen the relative wind will be parallel and opposite to the actual flight path regardless of the position of the nose in relation to the horizon. The rate of climb is a function of power and is determined by the amount of power available in excess of that required for level flight.

## TURNS AND COORDINATED FLIGHT

When you pose the question to a pilot, "What makes the airplane turn?", you often get the answer, "The rudder." It ain't so.

The rudder does not cause the turn; rudder causes yaw. The rudder is used during the turn to compensate for adverse aileron drag as the airplane is banked.

To begin a turn the aileron control is deflected in the direction of the desired turn, for example, to the right. This causes the right aileron to move up reducing the lift of the right wing (or by

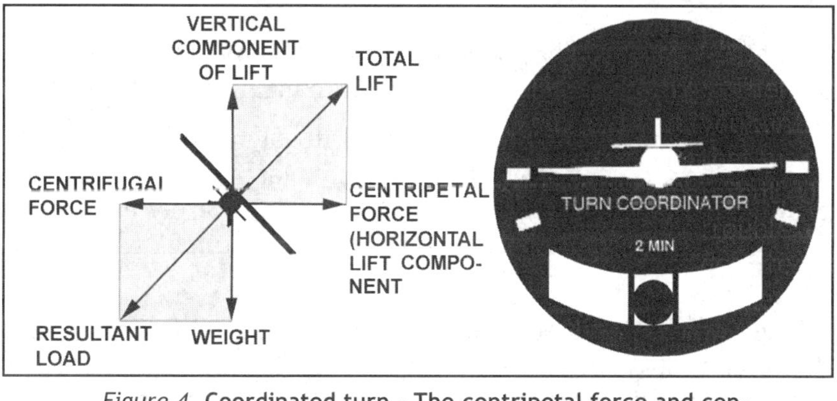


*Figure 4.* Coordinated turn - The centripetal force and centrifugal force are balanced.

creating camber in the opposite direction from what we normally think of as lift, it causes lift in the opposite direction). At the same time that the right aileron moves upward, the left aileron is deflected downward, increasing the lift on the left wing. This causes the airplane to roll to the right.

With lift and drag being directly proportional, when the lift is increased on the left wing, drag causes the nose to yaw to the left (during the right turn). The purpose of the rudder is to counteract this adverse aileron drag.

---

**To invent an airplane is nothing. To build one is something. To fly is everything.** – *Otto Lilienthal*

---

As you recall from your primary flight training, an airplane in level, coordinated flight may be banked to cause a turn. The turn is the result of a horizontal vector force working on the flight path.

In a coordinated turn the sideward portion of lift is called the horizontal lift component or *centripetal force*. Opposing centripetal force is *centrifugal force*. Notice the forces balance each other (figure 4) in a coordinated turn.

Suppose the airplane is banked to the right. Vector forces operate on the airplane.

The vertical force that opposes weight (gravity) sustains the airplane. The total lift must be increased to maintain level flight.

The horizontal force is called centrifugal force. "This is the force that an object moving along a circular path exerts on the body constraining the object and that acts outwardly away from the center of rotation." *Merriam-Webster's Collegiate Dictionary.*

Opposing the centrifugal force is the centripetal force. "This is the force that is necessary to keep an object moving in a circular path and that is directed inward toward the center of rotation." *Merriam-Webster's Collegiate Dictionary.*

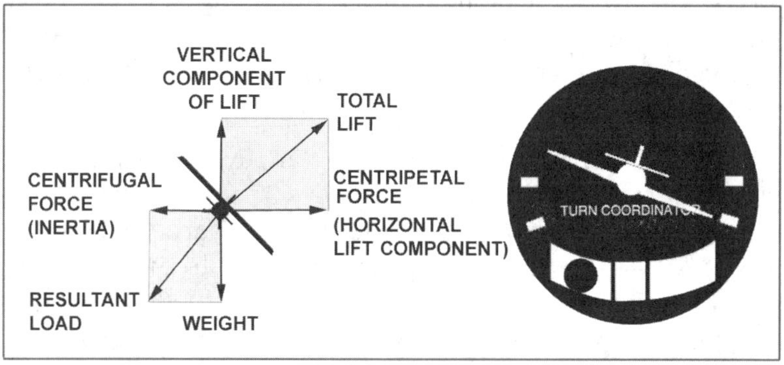


*Figure 5.* **SKIDDING TURN - The centrifugal force is greater than the centripetal force (horizontal lift component). The airplane moves to the outside of the turn.**

## Slips and Skids

In flight, during slips or skids, the flight path does not parallel the longitudinal axis of the airplane. An indication of the airplane's flight path will be shown by the ball of the turn-and-slip indicator, telling the pilot the "quality" of the turn.

During a skidding turn the centrifugal force is greater than the horizontal lift due to excessive rudder in the direction of the turn. Another way of looking at this is that insufficient aileron is being used. This causes the airplane to move to the outside of the turn.

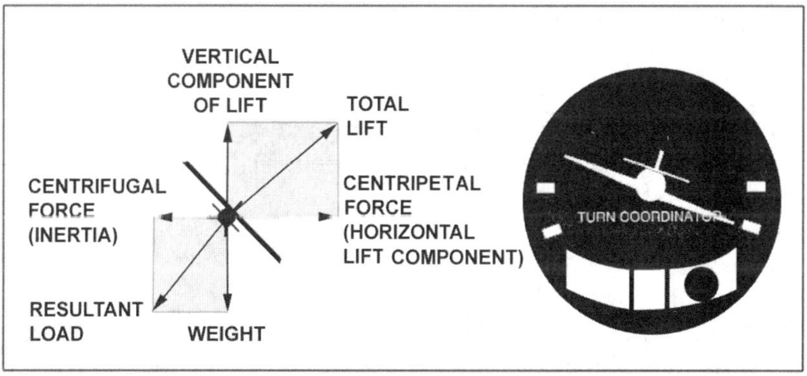


**Figure 6. SLIPPING TURN - The horizontal lift component (centripetal force) is greater than the centrifugal force component. This causes the airplane to move toward the inside of the turn.**

In contrast to the skidding turn is the slipping turn. The horizontal lift component (centripetal force) is greater than the opposing centrifugal force. This causes the airplane to move toward the inside of the turn.

This is an easy concept to prove. Maneuver the airplane to a ground reference maneuver altitude of 600- to 800-feet agl. Bank the airplane to the right and step a little too hard on the left rudder. The ball of the turn coordinator will move to the right indicating a slip. Being close to the ground you can actually see (and feel) the airplane move toward the inside of the turn. The airplane flies toward the ball.

Next, from the same turn, apply excessive right rudder pressure. As you do this the bank will increase, so adjust the ailerons as necessary to maintain the same bank angle. When the ball is to the left of the turn coordinator index look outside. The airplane is skidding toward the outside of the turn. You can see and feel this occurring. Again the airplane flies toward the ball.

REMEMBER: You might want to remember that the airplane always flies toward the ball of the turn coordinator.

## TWISTING FORCES

Airplanes having clockwise rotation of the propeller, when viewed from the cockpit, experience five twisting motions on the ground because of:

1. Torque reaction
2. Takeoff moment
3. P-factor or asymmetric loading of the propeller
4. Gyroscopic precession of the propeller
5. Corkscrewing effect of the propeller slipstream (spiral effect)

## TORQUE

Newton's Third Law of Physics (or Third Law of Motion)—every action has an equal and opposite reaction—explains the torque effect on an airplane. The propeller spins in a clockwise direction when viewed from the cockpit (in most airplanes). The equal and opposite reaction rotates the airplane in a counterclockwise

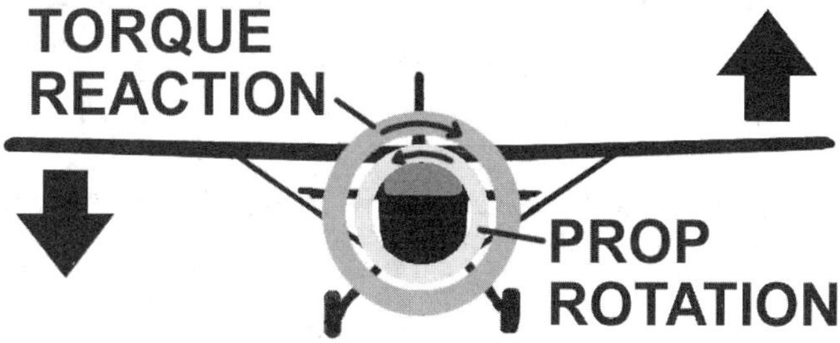


*Figure 7.* **Torque Reaction - Newton's Third Law of Physics states that every action has an equal and opposite reaction. The propeller revolves clockwise when viewed from the cockpit. The equal and opposite force tries to rotate the airplane counterclockwise.**

direction. Due to the differences in mass, the propeller cannot cause an equal and opposite rotation of the airplane; however, the force is strong enough to require aerodynamic modifications and pilot compensation.

To eliminate the need for pilot compensation of torque during cruise flight at cruise power settings, the manufacturer uses rigging (offset engine mounts, wash-in of the left wing, offset vertical stabilizer, or the like) to neutralize the torque.

Wash-in may be used on the left wing. Wash-in means the left wing is mounted at a higher angle of incidence than the right wing giving it more lift to counteract the torque. This additional lift causes the airplane to yaw to the left. To counterbalance the yaw the manufacturer may offset the forward portion of the vertical stabilizer to compensate for the extra drag at cruise airspeed condition. This has the same effect as stepping on the right rudder.

During a climb, the offset vertical stabilizer is not sufficient to compensate for the torque developed at slow speed and additional right rudder pressure is required.

During descent the offset vertical stabilizer over-compensates, requiring the pilot to make adjustments using the left rudder. If the descent is made with the power-off, the torque is reduced, requiring the pilot to compensate with left rudder. During a power-on descent, the airspeed is faster than normal, giving the offset vertical stabilizer too much authority. This also requires the pilot to use left rudder.

## TAKEOFF MOMENT

During the takeoff roll, as power is increased, torque is increased. The torque rotation of the airplane about the longitudinal axis (roll axis) causes the left wing to descend and the right wing to rise. This downward force of the left wing places more weight on the left-main tire, resulting in friction. The increased drag of the left tire creates a left-turning moment.

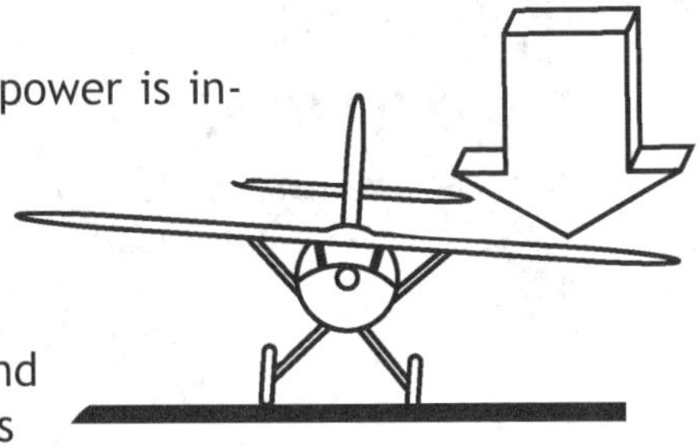

*Figure 8.* **Torque causes a drag increase during takeoff because the left tire is forced downward.**

This left-turning tendency becomes more noticeable in aircraft with powerful engines and "climb" props. The terrain of the operating surface—dirt, grass, sand, gravel, pavement—affects the amount of left-turning moment, varying from surface to surface.

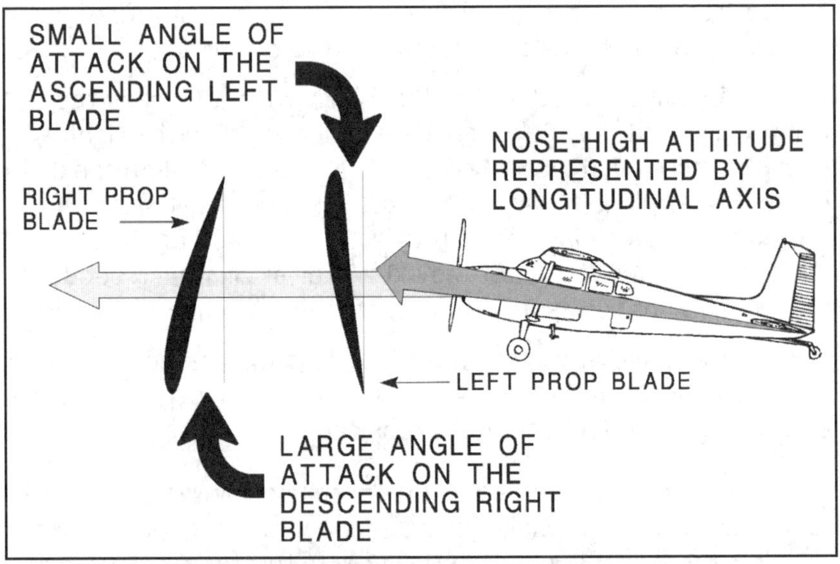


*Figure 9.* **P-factor - A nose-high attitude while the aircraft moves along the ground causes a left-turning tendency due to greater thrust from the descending right side blade of the propeller. During climbing flight the same thing happens. If the airplane was able to climb at an angle represented by the longitudinal axis, then the right propeller blade would have the same thrust as the left propeller blade.**

## P-FACTOR

P-factor or the asymmetric loading of the propeller results in a left-turning tendency of the airplane because the right side of the propeller produces greater thrust than the left side.

To simplify a complicated subject—where the angle of attack of both the airplane and the propeller must be taken into account—visualize an airplane climbing at climb power setting. This airplane has a certain angle of attack to produce lift (the angle between the chord line of the wing and the average relative wind). The flight path of the airplane does not parallel the angle of the longitudinal axis because there is insufficient thrust available for

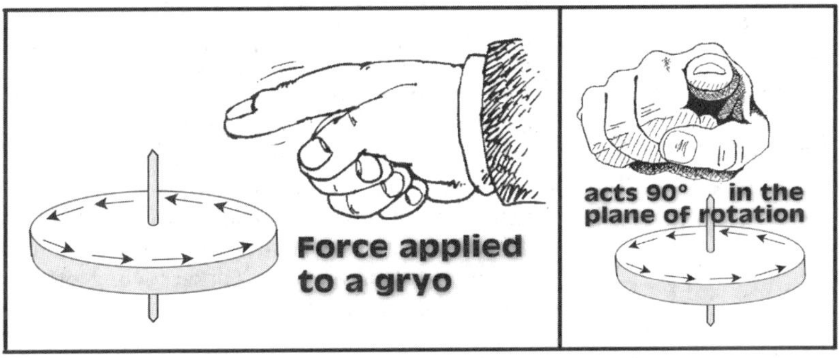


*Figure 10.* **The gyroscope's *precession* property means that a deflecting force applied to the gyro will cause a resulting force that acts 90 degrees in the plane of rotation.**

this to occur. The propeller blades are perpendicular to the longitudinal axis and point above the actual flight path. This results in the left ascending propeller blade having a smaller angle of attack than the right descending propeller blade.

## GYROSCOPIC PRECESSION

Gyroscopes exhibit two principles, ridigity in space and gyroscopic precession.

> *RIGIDITY IN SPACE - means that once a gyro is spinning, it tends to remain in a fixed position and resists external forces applied to it.*
>
> *GYROSCOPIC PRECESSION - An inherent quality of rotating bodies, which causes an applied force to*

*be manifested 90 degrees in the direction of rotation from the point where the force is applied.*

Any spinning object with mass will exhibit the properties of a gyroscope. One of the gyro's properties is called "precession." A gyro will tend to resist an applied force, but it will react in relation to that force. The reaction will be 90 degrees in the plane of rotation.

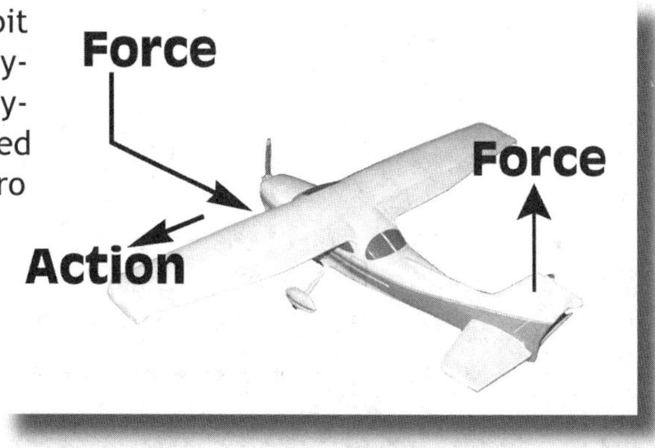


*Figure 11.* **When the tail is lifted abruptly during the takeoff roll, the lifting force would be the same as exerting a force to the top of the propeller arc. That force acts 90 degrees in the plane of rotation, creating a significant left turning tendency that combines with the other left-turn forces (torque, p-factor and the corkscrewing effect of the propeller slipstream).**

Gyroscopic action, as a result of the precession of the spinning propeller, will be greatest during takeoff when the tail is lifted from the runway. At this time, gyro precession causes a turning moment of the airplane to the left. The faster the tail is raised, the greater the left-turn force. Once the airplane is established in the takeoff attitude, the need for right rudder application disappears.

**There is only one slight difference between the tricycle-gear airplane and the conventional-gear airplane. The taildragger airplane wants to go down the runway backwards.** — *Sparky Imeson*

For visualization purposes, the force may be viewed from in front of or behind the propeller. It doesn't matter.

Raising the tail has the same effect as applying a force from the front of the airplane at the bottom of the rotating propeller's arc. The reaction will be 90 degrees in the plane of rotation, causing a turning moment to the left.

Viewed from behind, the force would be applied to the top of the propeller's arc, and moving 90 degrees to its right (the left when viewed from behind), the force would move outward on the left propeller blade to cause the left turn tendency.

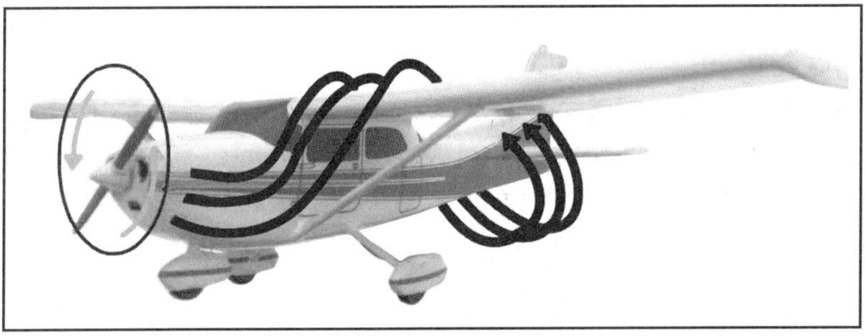

*Figure 12.* **The spiral slipstream (corkscrewing effect of the propeller slipstream) is concentrated on the left side of the airplane. If its arc is large enough to miss the left side, it also misses the right side.**

To reduce the amount of gyroscopic action, do not lift the tail abruptly during the takeoff roll. Raising the tail gently will spread the effect of the gyro action over a longer period of time, requiring smaller rudder inputs to correct any swerve tendency.

## CORKSCREWING EFFECT OF THE PROPELLER SLIPSTREAM

As the propeller rotates in a clockwise direction, it directs a stream of air rearward. This aft moving slipstream rotates in a corkscrewing manner. At high speeds it is elongated. During takeoff or slow flight it becomes more compact.

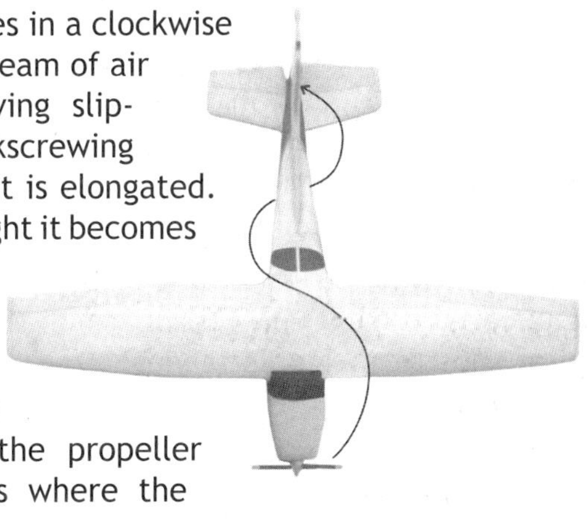

*Figure 13.* **The spiral slipstream (corkscrewing effect) is concentrated on the left side of the airplane.**

The corkscrewing rotation is clockwise when viewed from the cockpit, the same as the propeller rotation. In those areas where the corkscrew arc is of a large enough diameter to miss the left side of the airplane, it also misses the right side. When it's arc is small enough to contact the left side of the airplane, it is blocked from striking the right side of the airplane. This causes the majority of the corkscrewing effect to be concentrated on the left side of the airplane. One area where its effects are concentrated is the empennage, that is, the vertical stabilizer and rudder area.

When the corkscrew strikes the vertical stabilizer and rudder, on the left side, it creates a left turning moment. The slower the airplane's movement, the greater is this force.

## COMBINED EFFECT

The effect of all these forces combined requires the pilot to compensate with the application of the rudder. Not all these forces are present at the same time, and they vary in magnitude based upon the phase of flight, aircraft speed and pilot technique.

## STALL AERODYNAMICS

In most American-built airplanes, when climbing with the power on, right rudder application is required to compensate for torque, p-factor (asymmetrical thrust of the propeller), gyroscopic action, and the corkscrewing effect of the propeller slipstream (spiral slipstream).

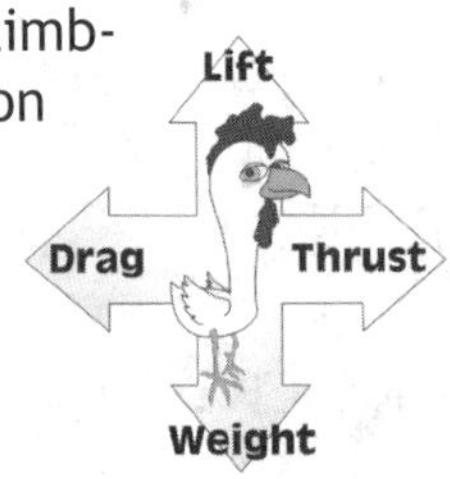

*Figure 14.*

Some people are not comfortable with slow flight and stalls. This is perhaps due to a misunderstanding of the aerodynamics involved during the stall. The following will eliminate the unknown from stalls.

Let's examine the effect that torque, p-factor (asymmetrical thrust of the propeller), gyroscopic precession, and the spiral slipstream (corkscrewing effect of the propeller slipstream) exert on the airplane during stalls.

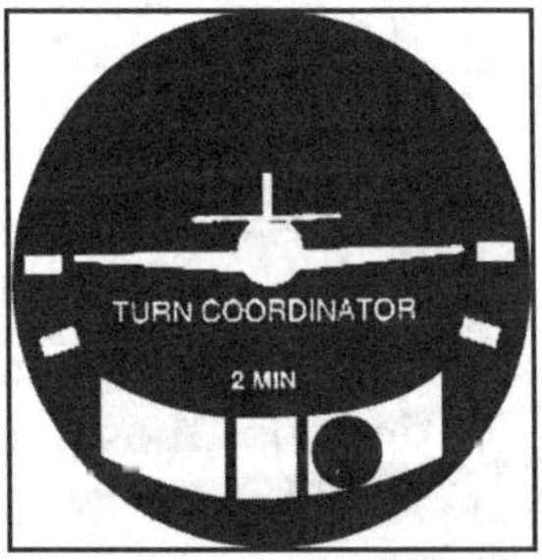

*Figure 15.*

Recall that the relative wind is created by the flight path (parallel and opposite) of the airplane. During slips and skids the flight path is not parallel to the longitudinal axis.

Establish a power-on climb. If all these forces are not compensated for, the ball of the turn-and-slip instrument is to the right of the index (*Figure 15*). **The airplane flies toward the ball.** If we continue the climb to a stall, a portion of the airflow to the left wing is blocked (low-wing or high-wing airplane). If the airplane stalls, the left wing will stall first, rotating in a direction opposite the ball.

Flight instructors need to realize, during flight instruction, that it is not a good idea to say, "Now watch this, the left wing is going to fall off." This miscommunication can really scare a student. It's more comforting to say, "The airplane will rotate to the left because the airflow is blocked to the left wing, causing it to stall first."

---

**I ask people who don't fly, "How can you not fly when you live in a time in history when you can fly?"** — *William Langewische, 2001*

---

The airplane creates the relative wind, relative to the direction the airplane is moving. When the ball is to the right of the index marks, it shows that the airplane is flying toward the ball. The relative wind is parallel and opposite the flight path. The fuselage blocks part of the airflow to the left wing, causing it to stall first, with the resultant rotation away from the ball.

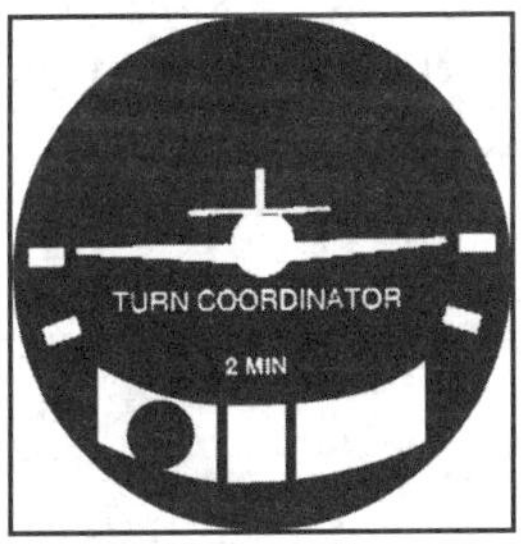


*Figure 16.*

If, during the climb, the right rudder application is too hard, the ball moves to the left (*Figure 16*). The relative wind now approaches the airplane from the left. That portion of the relative wind that should flow over the right wing becomes blocked by the fuselage. The airplane, if stalled, will rotate to the right away from the ball.

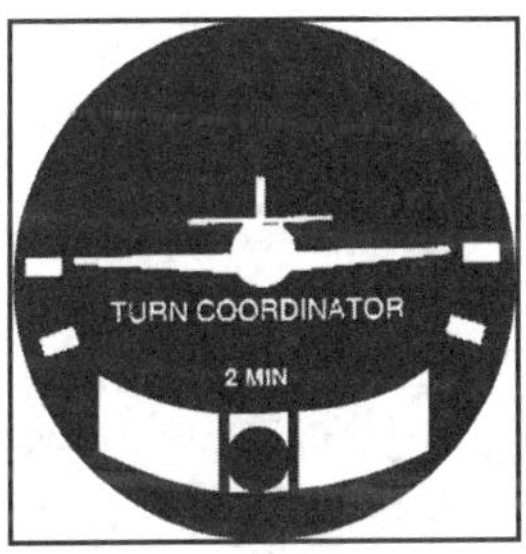


*Figure 17.*

By learning that the airplane flies towards the ball, you can predict what the airplane is going to do in a stall situation.

Thinking we know all there is to know about stalls, we begin a power-on climb and center the ball of the turn-and-slip indicator (*Figure 17*). We are very careful to keep the ball centered, knowing if the ball is to the right, the airplane will rotate away from the ball, twisting to the left. If the ball is to the left, the right wing stalls first. So at the moment of stall, with the ball centered, the darn airplane, unexpectedly, falls to the right.

Why? When the angle of attack increases, the effect of the spiral slipstream moves up the vertical stabilizer and rudder.

During the climb, to keep the ball of the turn coordinator centered, right rudder application is used to compensate for the four forces. As the stall is approached one of the forces (the spiral slipstream) is disappearing. There is too much right rudder and at the stall the airplane rotates to the right.

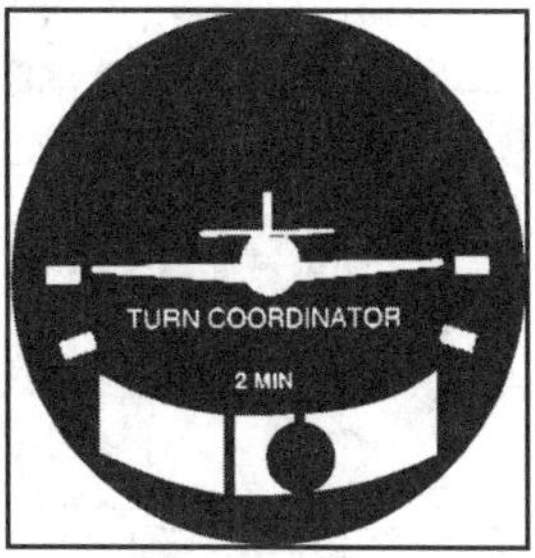

*Figure 18.*

If it is desired to practice power-on stalls and have the airplane fall through the horizon with the wings level, keep the ball halfway out of the index marker to the right (*Figure 18* - half in and half out).

---

**We contrive to make the invisible air support us, we relinquish the security of feet on the ground because flying is demanding, delightful, beautiful; because we love it. Very few of us are actually crazy, and nearly all of us manage the risks as well as we can, but we all willingly trade some of our security for the immeasurable beauty of the sky.** — *Paul J. Sampson*

---

This understanding, developed by knowledge of aerodynamic forces, will aid in reducing the apprehension associated with stalls and slow flight maneuvering. You are able, beforehand, to predict what the airplane is going to do and why it is doing it.

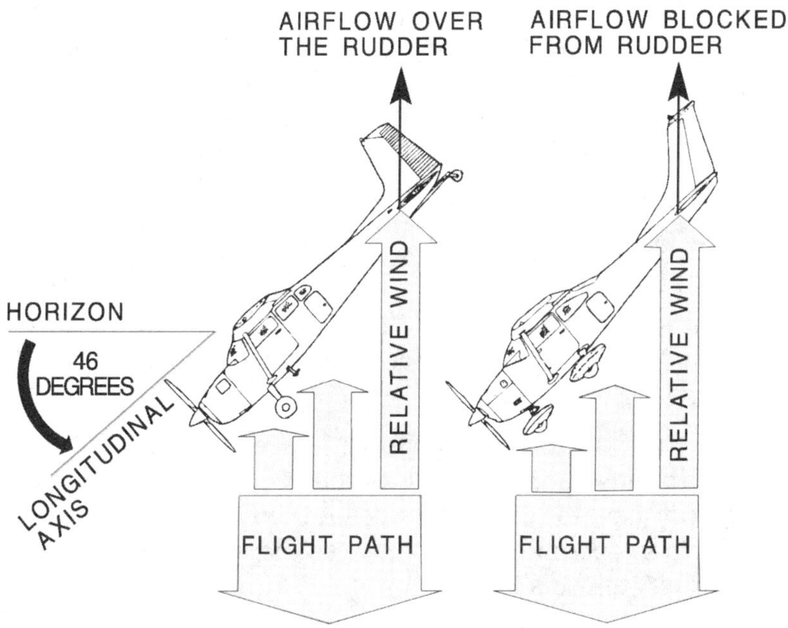


*Figure 19.* Straight-tail airplanes have rudder control during a spin. Swept-tail airplanes require forward movement of the control wheel (stick) to force airflow over the rudder before the rudder is effective in stopping the rotation of the spin.

## STRAIGHT-TAIL VERSUS SWEPT-TAIL AIRPLANES

Somewhere around 1960, aircraft manufacturers switched from the straight tail to the swept-tail design. This was an important sales tool. Just look at the airplanes on any ramp and compare the straight-tail models versus the swept-tail models. The swept-tail airplanes look good, in fact, it looks like they are going 15 mph just sitting there.

An intentional or unintentional spin requires that you realize the swept-tail airplane incorporates a different spin-recovery technique from the straight-tail airplane.

*Figure 19* shows that the airflow to the rudder of the swept-tail airplane is blocked by the horizontal stabilizer/elevator.

### Straight-tail Spin Recovery

The straight-tail airplane incorporates the traditional spin recovery technique; step on the opposite rudder to stop the rotation, then stick forward to break the stall (or both at the same time, and stick back to recover from the resulting dive.

### Swept-tail Spin Recovery

Many airplanes assume a nose-down attitude of approximately 45-degrees below the horizon while in a spin. The uninitiated swear the nose is pointing straight down. Despite the attitude of the airplane, the flight path is downward. The resultant relative wind is upward.

During a spin recovery, if opposite rudder is used to stop the rotation, nothing happens. The horizontal stabilizer and elevator block the air flow to the rudder because of the swept-tail design. Recovery in swept-tail designs requires the stick (control wheel) be moved forward to allow air flow over the rudder. Only then will rudder application stop rotation.

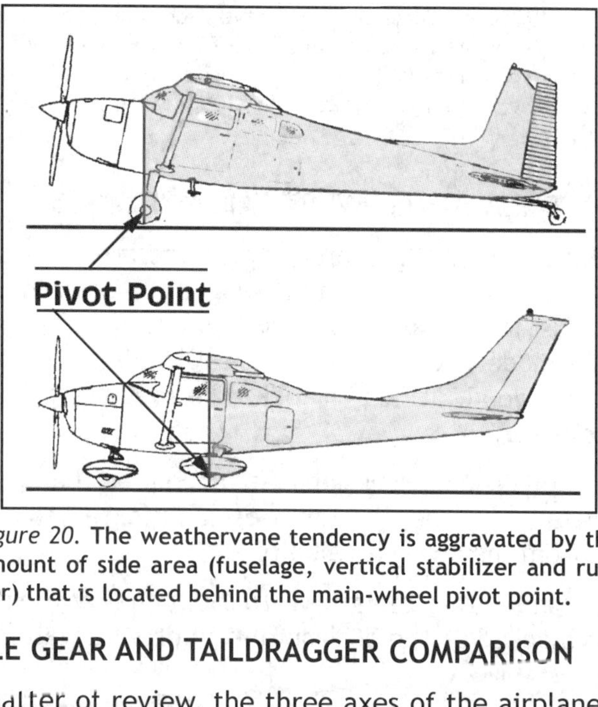


*Figure 20.* **The weathervane tendency is aggravated by the amount of side area (fuselage, vertical stabilizer and rudder) that is located behind the main-wheel pivot point.**

## TRICYCLE GEAR AND TAILDRAGGER COMPARISON

As a matter of review, the three axes of the airplane are perpendicular to each other. The vertical axis (yaw), lateral axis (pitch) and longitudinal axis (roll) all pass through the center of gravity (cg). The cg is considered as that point where all the weight of the airplane is distributed and balanced. Movement of the airplane occurs about the cg.

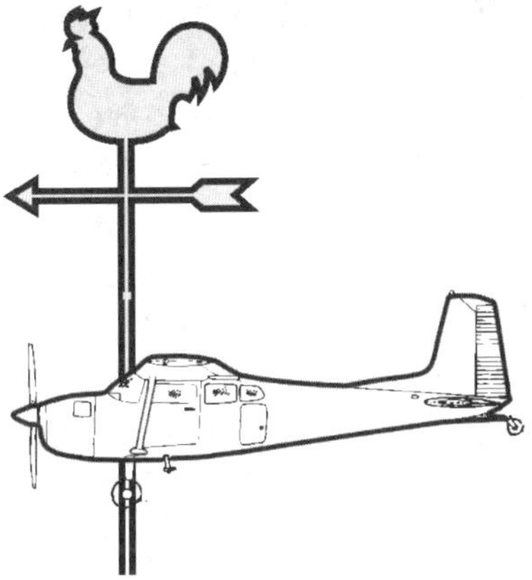

*Figure 21.* **Weathervane**

The stability of any particular airplane determines its flight characteristics. The addition of wings, horizontal stabilizer and rudder enhance the stability. If an outside force causes the airplane to begin a turn without control application by the pilot, the relative wind striking the side of the fuselage and rudder will straighten the flight path. If the aircraft pitches up or down due to turbulence, the horizontal stabilizer and elevator will bring it back to the original flight path. In this manner, the controls act like a weathervane in maintaining the airplane's flight path. In flight, both the tricycle gear airplane and the tail wheel airplane have similar stability.

On the ground, the similarity between control of tricycle and tail wheel airplanes assumes a different dimension. The weathervane tendency of a taildragger airplane is due to the side area exposed behind the main-wheels-pivot point.

There is more side area exposed on the taildragger. With the main wheels acting as a pivot point and a greater profile behind the main wheels, the airplane naturally wants to turn or weathervane into the wind. There is no nose wheel to retard or stop the weather vane tendency.

## CENTRIFUGAL FORCE

Centrifugal force increases as the square of the ratio of the actual speed where it starts divided by the desired speed. Any swerve, during takeoff or landing, will be worse as the speed increases. Even a small change in groundspeed at touchdown may affect the outcome of a swerve.

For example, while landing, if the groundspeed is doubled, this means any swerve will result in quadrupling the centrifugal force involved. This force may exceed the capability of the rudder and brakes to counteract the force. In such a situation, full power application will help straighten the airplane by pulling it along the thrust line and by providing propeller blast on the vertical stabilizer and rudder, to make them more effective.

The normal three-point landing (also called the full-stall landing) is accomplished at the slowest speed possible. Once on the ground the stick is pulled full aft to enhance tail wheel steering. This makes control during the after-landing roll easier for the pilot.

---

> **I learned to watch, to put my trust in other hands than mine. I learned to wander. I learned what every dreaming child needs to know — that no horizon is so far that you cannot get above it or beyond it. These I learned at once. But most things come harder.** — *Beryl Markham, "West With The Night,' 1942*

---

## SUMMARY

Negative transfer, similar to controlling an automobile with the steering wheel, may cause the pilot to try to control his airplane with the control wheel. In a situation where the airplane is slow (near its critical angle of attack) and a wing drops down, full aileron deflection will not raise it. Rudder application, however, corrects the situation immediately.

With an understanding of the twisting forces of flight comes the realization of the importance of the rudder in maintaining safe flight, on the ground and in the air.

Additional aerodynamic forces will be explained in the Taxi, Takeoff, Landing, and Crosswind Operations chapters.

*Figure 22.* **Maule M-4s in formation flight.**

# CHAPTER 2

# Preflight and Starting

## PREFLIGHT

When a pilot gains proficiency and becomes comfortable with taildragger (or tricycle gear) operations, there is a tendency to abbreviate or sometimes to completely skip the preflight. The onset of indifference about performing a proper preflight is your warning sign of complacency.

This is especially true when the temperature drops and the weather turns lousy—just when the preflight is needed the most. Dress appropriately for the conditions, and take your time checking everything on the airplane's exterior and interior.

*Figure 1.* Perform a thorough preflight inspection before beginning any flight. Remember: "In God we Trust ... everything else we check."

## WHAT TO CHECK

The taildragger preflight is similar to the tricycle-gear airplane. Follow the same steps and keep in mind that the landing gear and drain holes on the bottom of fabric planes need special attention. Plugged drains may cause corrosion.

## LANDING GEAR

Three types of fixed-landing gear are found on conventional gear airplanes (but not on the same airplane).

**Shock cord (bungee rubber cord)** - Airplanes incorporating the shock cord landing gear have steel tubing mounted in a manner to produce a lever action for the gear struts. Bungee cords are wrapped around the tubing to act as a shock absorber. Bungee cords are color coded in 5-year cycles to allow mechanics to quickly check the age of the cords.

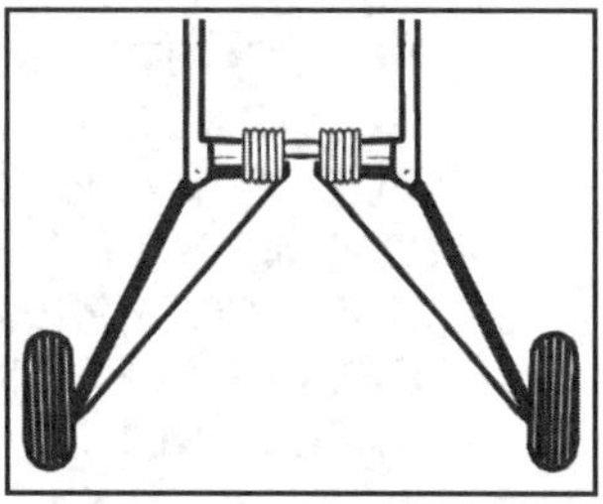

*Figure 2.* **Shock cord**

**Spring steel** - The gear is made of a tapered strip of strong spring steel. The top (wide end) is attached to a gear box and the narrow end has an axle attached for the wheel.

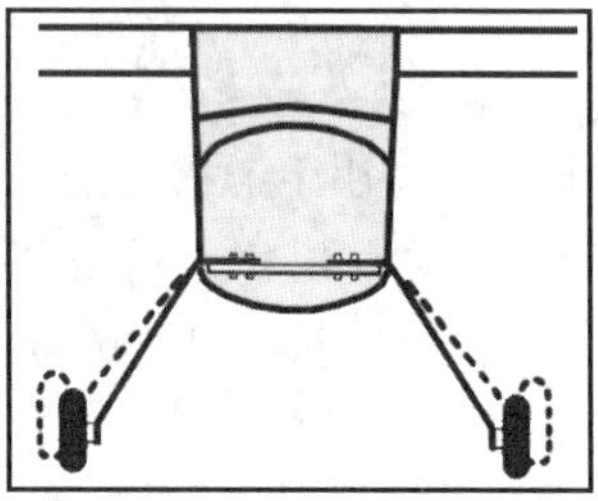

*Figure 3.* **Spring steel**

**Air-oleo shock struts** - This works like a shock absorber on a car. Within the housing is a piston that moves up and down. It is filled with fluid and the dampening occurs when the fluid is forced through an escape valve (metering pin).

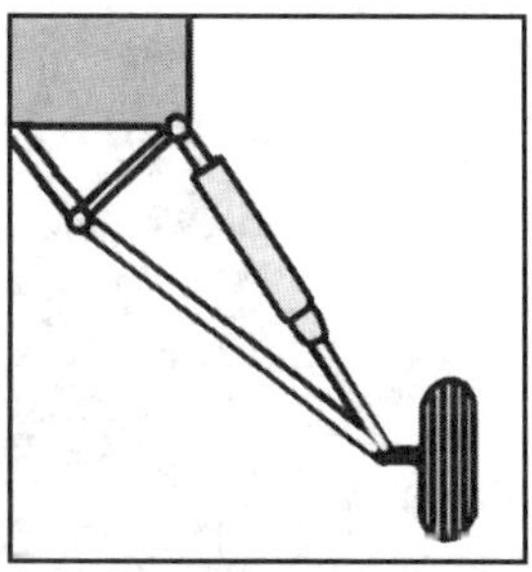

*Figure 4.* **Tripod-oleo**

Regardless of the gear type, check the fuselage attach points for play, that is, any movement that might indicate a problem. If the wing tip is moved up and down, the gear should flex. If the wing tip is moved forward and aft, there should be no movement of the gear arm.

## TIRES

Proper tire inflation ranks as the most overlooked and easiest to solve maintenance function that can be performed by the pilot. Keep in mind that when a new tire is installed it will 'grow' for about a day. As it expands the pressure will decrease. So checking the inflation each day for a couple of days is a good operating practice.

Most tire manufacturer's recommendation to check the tire inflation every 24 hours or before every flight of infrequently flown airplanes may seem excessively burdensome. Most pilots work out some general rule such that after checking the tires for a couple of days and the pressure remains constant, then going a week or two between checks does not negate his preventive maintenance program.

When checking the tire pressure, avoid overinflation that will cause excessive wear along the center. Underinflation is more often the case, and it causes the outside edges (shoulder area) of the tire to wear abnormally and usually leads to tire damage, wheel damage and vibration problems.

The aircraft maintenance manual will provide information about the proper inflation and more importantly, the type of tires that may be installed on a particular airframe. The airplane specifications may require a 4-ply tire with a pressure of 29 psi (pounds-per-square inch). If a 6-ply tire is installed with a pressure of 42 psi, when you (or another pilot) check the POH (Pilot's Operating Handbook) or Owner's Manual for inflation information, this tire (in accordance with the POH recommendation) will be underinflated by 13 psi, leading to tire damage and vibration.

Occasionally you will find a difference between the tire manufacturer's recommended pressure for a particular tire and the aircraft manufacturer's recommendation. This may be due to airframe requirements (especially nose wheels) and certification. In this case, use the aircraft manufacturer's specification.

If a tire has experienced a large pressure loss of 80-90 percent of its recommended pressure and, for whatever reason, it is operated for taxi, take off or landing, the tire should be replaced.

## FUEL CONTAMINATION

Fuel contamination, consisting mostly of water, is always a possibility in aircraft fuel systems. It does not matter whether the airplane is parked on the ramp, or tucked inside a cozy hangar. It is still susceptible to condensation of water.

Contamination is also possible during cold weather. Some pilots feel that cold air does not contain sufficient moisture to enable the formation of water in the fuel. For the most part this is true, but there are those days (and nights) when the air will condense water into the fuel supply. Not knowing exactly what day that will be, checking each individual tank sump, cross-feed drain, and main sump or gascolator is best. If a sump refuses to drain, find out why. It may be contamination or frozen water.

Taildraggers afford access to remote airstrips and the possibility of using makeshift-fueling facilities. Fuel drums, even if refinery sealed, may contain rust and other contaminants.

Avoid using any gas other than the proper grade of aviation fuel (or autogas STC) specified for your airplane. If the proper grade is not available, check the pilot's operating handbook for suitable and acceptable substitutes. At a makeshift-fuel facility, be certain to filter the fuel, use a static-ground line, do not use a plastic funnel, and do not wear nylon clothing while refueling. Fuel flowing through a plastic funnel can create static electricity. Nylon causes static electricity.

## AUTOGAS STC

What in the devil is an STC?

"STC" is an acronym for Supplemental Type Certificate. The STC is an FAA-approved engineering data package that provides details of a specific installation. STCs are required for many installations to certify that the alteration is in accordance with FAA requirements.

Auto fuel STCs deal with particular engine models and specific airframes. The issuer of the STC has gone through a lengthy certification procedure testing for detonation, vapor lock, endurance, and the like.

You shouldn't figure that just because your friend has an STC for his airplane and your airplane is the same make and model, that you can go ahead and use auto fuel without shelling out the bucks for the approval. This would be a violation of FARs. And, your insurance may be void if you do so.

Why should you even consider an auto fuel STC? The most lucrative reason is cost savings. The Experimental Aircraft Association's (EAA) research shows the average recreational pilot flies about 70 hours per year using an average of about 10 gallons per hour. With a savings of 50 cents (minimum) per gallon, this means an extra $350 in pocket.

In addition to cost savings on fuel, there are claims of reduced maintenance costs due to fouled plugs and sticking valves.

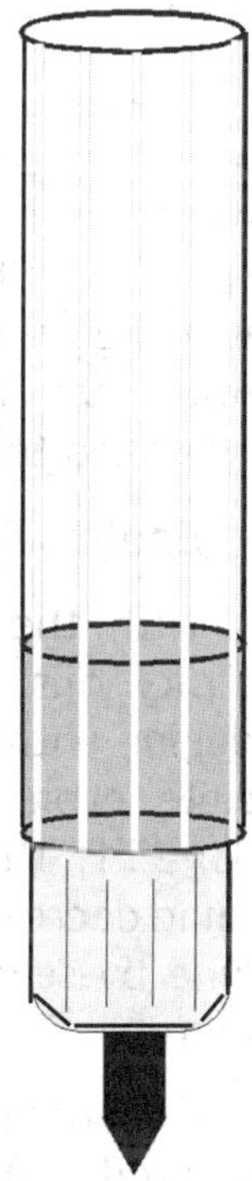

*Figure 5.*

As far as safety, autogas has been used for more than 20 years. The FAA issued a Safety Program Newsletter in early 1998 that contained information inferring that airplanes and engines with an STC approval for autogas are not as safe as airplanes using avgas exclusively. On June 4, 1998, the FAA issued an apology to the EAA, stating that autogas is an acceptable alternative to avgas.

## FUEL SAMPLING PROCEDURE

Regardless of the type fuel you use, for the first flight of the day and after each refueling, draw off sufficient fuel into a transparent container to see that it is free of contaminants.

If the airplane is moved from a hangar, the fuel will probably slosh against the sides of the tanks and wash down any water present causing it to go into suspension with the fuel. Place the airplane in a wings-level position. Allow at least 10 minutes for the fuel and water to settle and the water to separate, then drain the sumps and filters.

The proper technique for obtaining the fuel sample is not to drain a container full of fuel all at once. In the northern hemisphere, a low pressure area results in a counterclockwise rotation of a fluid. When the sump is drained by holding the quick-drain valve open, a low pressure is formed that causes water and contaminants to be swirled to the outside of the miniature tornado inside the tank. By taking small bursts from the sump, you eliminate the fluid swirl that slings water and contaminants away from the drain. If the screwdriver-type fuel sampler is used (Figure 5), each burst will fill the tube about one-fifth to one-quarter of its capacity.

## ENGINE STARTING

Proper starting procedures will prolong the life of the engine and provide more reliable operation. It is said that 90 percent of the engine wear occurs during the first 10 seconds after the start.

The reliability of the modern aircraft engine is good. Compared with the reliability of turboprop or turbojet engines, the modern reciprocating engine contains the warning, "it's not if the engine is going to fail, it's where it will fail." Don't cause unnecessary wear and tear on your engine.

### Engine Preheat

With the advent of multi-viscosity oils, starting an airplane engine without preheating is possible, even in temperatures as low as -25°F. That is, if one does not mind decreasing the life of the engine by several hundred hours or more.

Generally, +20°F or lower is considered "preheat" temperature. Below the pour point, oil quickly

*Figure 6.* **Frosted sparkplugs can turn even the nicest guy into an ogre. Frosted plugs require heat, either engine preheat or removal of the sparkplugs. Relax with a cup of coffee while the engine is being warmed.**

turns to a molasses-type liquid and any colder temperature results in the oil rapidly approaching a solid.

## Engine Starts

Above +20°F, engines may be started without preheat, but the oil may be partially congealed. During "chilly" temperature conditions, turn the propeller over by hand at least four times to relieve some stress on the starter and battery. Remember a broken p-lead (ground wire) on a magneto may create a "hot" magneto even when the magneto switch is in the off position.

In the "old days" we turned the prop backward because a hot mag will not fire when turned in this direction. Check with a mechanic before turning the propeller backwards on modern engines. Some part of the engine may be damaged, e.g., vacuum pump vanes may chip and break on some engines.

Never operate a starter motor for more than 90 seconds continuously. If the engine does not start, allow a full minute before attempting another start to allow the starter motor to cool down. After the initial attempt, do not operate the starter for more than 30 seconds continuously. (Check the manufacturer's recommendation for starter operation. The continuous starter operation may be 30 seconds with a four-minute cool down between start attempts, or some other such values.)

During cold weather, an engine that has not been preheated is susceptible to "frosted plugs." The engine fires a couple of times and then quits. Any further attempt at starting is futile because there has been sufficient combustion to cause some moisture in the cylinders, but insufficient combustion to heat the spark plugs. The water condenses on the spark plug electrodes, freezes to ice, and shorts them out. The only remedy is heat. If an engine heater is not available, remove the plugs and heat them until all moisture disappears.

For prolonged cold weather operations, switching to fine-wire spark plugs helps eliminate the frosted plug problem and improves cold starts.

## Priming

Sometimes pilots have a tendency to over prime during cold starts. This produces the undesirable effect of washing the oil from the cylinder walls. Priming causes raw fuel to be directed into the cylinders without passing through the carburetor and intake manifold. After starting, lack of lubrication causes scoring of the cylinder walls. Once this occurs, the resultant lack of engine compression causes hard starting.

Over priming can also cause induction fires. If a fire occurs, continue operating the starter in an attempt to suck the fire back into the engine where it belongs.

### Throttle Usage

Immediately after the engine starts, do not operate above 800 to 1,200 rpm, until several minutes of running loosens (warms) the oil to provide lubrication. Revving the engine as little as 1,500 rpm immediately after starting will reduce the life of the engine.

In an attempt to speed up heating of the engine after the start, some pilots have closed the cowl flaps. They find out later this may develop "hot spots" that have the potential of damaging engine accessories, wiring and supply lines.

### Propeller Over Dirt or Water

Once an engine has started, the spinning propeller creates a low pressure area. Whenever the airplane is in an area of dirt, rocks or pooled water, do not use high power settings while parked or beginning to taxi. Manually position the airplane away from pools of water.

When power is added, small objects can be picked up and pulled into the propeller's arc. The impact of sand, dirt, gravel or even small droplets of water funneled to the prop can turn the blades into something resembling a chain saw blade. Small water droplets may be fully as destructive as rocks when hit by the rapidly rotating propeller blades.

### Beginning to Taxi

When beginning to taxi and the airplane seems like it will not budge, pump the control wheel fore-and-aft while adding power. If this does not cause the airplane to "break loose," push the rudder back and forth and rock the ailerons. This technique will allow the airplane to begin movement with less power application.

### Oil Pressure Indication

Normal starts should produce an oil-pressure indication within 30 seconds after starting. During cold temperatures, there should be an indication within 60 seconds after starting.

If there is not an indication of oil pressure, turn off the mags, rather than the mixture, to stop the engine. This will result in immediate stopping. If there is a lack of oil pressure, there is less chance of engine damage when the engine is shut down immediately. Investigate the problem.

Remember that by turning the engine off with the magnetos, unburned fuel may remain in the cylinders. If there is a glowing piece of carbon on the piston head, movement of the propeller may produce enough compression to cause the engine to fire over a time or two with the potential of harming the person handling the prop.

The engine may be receiving adequate lubrication and the problem is the result of congealed oil in the transmitter line to the oil pressure indicator. Then, again, there may be a problem that is preventing the engine's life blood from circulating.

### Carburetor Heat

During cold weather, using carburetor heat to help atomize the fuel is occasionally necessary so the engine will run smoothly. When using carb heat, the normally filtered air is shunted through a shroud so that unfiltered air is taken directly to the carburetor. If the airplane is parked in an area of dust or sand, this can lead to foreign object ingestion inside the cylinders.

## HAND PROPPING

On a beautiful spring day during my primary flight instruction, Rich Herman and I were taking turns practicing accuracy landings at Columbus, Montana. It was my turn to stand beside the runway and mark Rich's touchdown point. A Beech Debonair landed and deplaned two elderly passengers. The pilot could not restart the engine. He got out and prepared to prop the plane. I yelled down the field, asking if he wanted some help. He said, "No," and I watched the procedure.

The engine started with the throttle at full power. The surprised pilot stumbled backwards and fell down. The airplane began moving and passed over the frightened pilot. It first crossed the runway, headed for the corner of a baseball diamond to the southwest, where the left wing broke the 30-foot-high pole near home plate during impact. The airplane swerved left, lurched through a ditch and pitched across the street toward a small white house. It was stopped, just before becoming an uninvited luncheon guest, by a big cottonwood tree.

Luckily, property damage was the only consequence of this accident. Hand-propping accidents are likely as not to cause serious or fatal injuries.

Back in the "good old days," all airplanes were started by someone swinging the prop. Many modern-day aviators have not been exposed to the procedures and precautions necessary to safely start the airplane engine by hand-propping. Yet, with the increasing popularity of taildragger operations over the last decade, more pilots are being exposed to airplanes without an electric system and consequently, hand propping.

Before participating as a substitute electrical system and starter, you better know the rules of the game, for it is a dangerous one.

Hand-propping, even when properly done, is still a game of chance. Most, but not all, of the risk can be eliminated by adhering to established rules.

### Initial Preparation

Before doing anything, check that the brakes are operating and set. Besides setting the brakes, the safety pilot should hold the brakes. Chock the main wheels.

If you are propping the airplane by yourself and there is a glider tow hook on the airplane, connect the tail tie-down rope to the hook. After starting the engine, jump in and pull the handle to release the tie-down to be on your way. Unless you are certain the brakes are set and will hold once the engine is started, it's a good idea to use the tail tie-down even without a tow hook.

When propping without assistance, it is prudent to turn the fuel selector "off." Experiment with the airplane you fly. With the fuel selector turned off, the engine should continue to run a minimum of 30 seconds to one minute before running out of fuel.

*Figure 7.* **Standing too close while swinging the prop prevents an escape from the propeller's arc.**

*Figure 8.* **Standing back too far causes the pilot to fall into the propeller's arc after the swing.**

## Precautions

When propping from in front of the airplane, a proper stance is important. Standing too close prevents escape from the propeller arc once the engine fires. Standing too far away can cause you to fall into the propeller arc.

A position where you lean forward slightly, with the weight on the forward foot, allows body movement backward during the swinging of the prop with weight transfer to the rear foot. This

*Figure 9.* **Proper Stance for "swinging the prop." Stand back far enough so that you have to lean forward to place the fingertips over the prop blade. When the prop is swung, your motion will carry you back and away from the prop. Remember, use the fingertips, not the knuckles, to grasp the propeller blade.**

posture causes the body weight to move away from the prop arc as the blade is swung.

With your left foot forward (carrying the majority of your weight), simultaneously pull down on the prop and push back with the left foot. Have a knowledgeable instructor (not necessarily a CFI) demonstrate this technique of swinging the prop; then have him do a critique while you try the maneuver.

### Mag Test

Pilots flying taildraggers that require hand-propping develop a habit of testing the mags upon shutdown to insure one or both of them have not become "hot." The test is performed at low rpm, around 660-700 rpm. The ignition is momentarily turned to the off position. The engine should stop firing and the rpm begins to drop. After a drop of 200-300 rpm, the ignition is turned on again. Doing this test at high rpm can cause engine damage.

A hot mag can occur when the "p-lead," a ground wire to the magneto, becomes loose, bare or broken. Also, a faulty ignition switch may create the "hot" condition. There are other possibilities, but all of the mechanical devices in an airplane have been designed to maintain a live condition with their failure. This "fail safe" design means that should any of these glitches occur, even with the mag switches in the "off" position, the mags are hot and any rotation of the propeller, even if it is done slowly by hand, can cause the engine to fire.

## PROPPING PROCEDURE

Familiarize the safety pilot with the controls, brakes and systems. Establish the commands to be used and what they mean to each person. When in position to prop the engine, with the throttle cracked, brakes set, and primer as required; yell, "Contact" or "Mags on." The safety pilot should switch the mags on and respond with, "Contact" or "Mags on."

The finger tips (never beyond the first joint) are placed on the propeller blade. It's more comfortable to grip the blade with the second knuckle of the fingers, but it is unacceptable. The position of the finger tips along the length of the propeller depends on the engine. Powerful engines require moving the finger tips toward the tip of the propeller to have enough leverage to spin the propeller. Most airplanes can be started with the fingertips placed about two-thirds of the way out from the hub or spinner.

Another position used on older taildraggers with less powerful engines is standing behind the prop on the right side. The left arm can hold the wing strut or door frame while swinging the prop with the right hand. A disadvantage of this technique is the requirement to wrap the fingertips over the trailing edge of the propeller. An advantage, when starting alone, is that it places the pilot closer to the cockpit controls.

The engines in most American-built airplanes have clockwise rotation when viewed from the cockpit. This obviously means the person doing the swinging of the prop, when facing the propeller, would move the propeller on his left side in a downward motion.

Sometimes the propeller blade ends in an awkward placement where propping is impossible. When repositioning this prop blade, assume the mags are hot and that the engine can fire.

Back in the days before engines were equipped with starters and fancy flywheel-gear connectors or direct-cranking starters with overrunning clutches, it would not hurt the engine to be turned backward with the propeller. When the engine is turned backward, the mags will not fire, even if hot. So, repositioning the propeller is easy by turning it backwards. Some newer engines can be harmed with backward rotation. Check with a mechanic to see if your particular engine can be rotated backward without causing harm.

## SAFETY RULES

- *Have a knowledgeable person (safety pilot or mechanic) in the cockpit.*
- *Hand-propping can be done without the safety pilot (mechanic); however, additional precautions must be taken.*
- *The ground around the propeller must be free of impediments, ice and snow.*

- *Remove all loose articles of clothing. Remove rings from the hand or hands used to swing the propeller.*

- *Stand close enough to the propeller to eliminate an off-balance condition from occurring when the propeller blade is swung.*

- *Use caution whenever handling or positioning the propeller. Assume it is "hot" whenever it is touched.*

- *When ready to swing the prop, talk to the cockpit safety pilot.*

Once the airplane is started, taxiing becomes the next challenge.

---

**An old, bold taildragger pilot is one who has had some interesting experiences ... some of which are true.** — *Sparky Imeson*

---

# CHAPTER 3

# Taxiing

## INTRODUCTION

Hangar flying, that area of aviation where one can sit around and visit (or swap lies) with other pilots about flying, is a great way to gain additional knowledge about procedures and techniques.

Hangar flying can provide valuable lessons ... until you encounter and believe someone who talks, not about their experiences, but about what they have heard to be true. Thus the beginning of many rumors concerning taildraggers.

Not to be outdone at the next hangar session, the stories are embellished until they are blown out of proportion.

> **A journey of a thousand miles begins with a single step.** — *Confucius*

The hard-to-handle reputation of many taildraggers is often exaggerated. For example, in comparing the Luscomb with the Aeronca you find the Luscomb stays on the runway for a longer takeoff distance. This requires, while maybe not greater concentration or skill by the pilot, at least concentration for a longer period of time. The Cessna 195 also has the reputation of being harder to handle than similar planes because of the extended time on the ground.

Although you should take these tall tales in stride, there must be some spark of truth in there somewhere. So regardless of the exaggerations, this type of information should serve as a warning to never let down your guard when flying a taildragger.

The taildragger presents the pilot with more of a challenge than flying an airplane with training wheels (tricycle gear). It requires knowledge and practice with some good old common sense thrown in. In other words, flying a taildragger is a full-time job requiring a full-time pilot.

## TAIL WHEELS

Tail wheel airplanes use various configurations for the tail wheel assembly. Depending on the make and model, you might find:

- *Skid - This is the configuration that created the name 'taildragger.' The skid is an effective brake when the stick is pulled back to your gut.*
- *Non-steerable (free caster or swiveling) - Control is with the use of the rudder and brakes.*
- *Steerable - There is an interconnect with the rudder so when the rudder is pushed, the tail wheel tracks to assist in the turn.*
- *Steerable locking - There is an interconnect with the rudders and a lock that can be engaged*

*Figure 1.* **Tail wheel assembly (Aviat Husky). Pulling the stick back to your gut makes steering more effective.**

*Figure 2.* **Pay particular attention to the springs during the walk-around inspection checking alignment and security.**

*in the cockpit prior to takeoff to cause the tail wheel to track straight.*

- *Steerable retractable - Retractable tail wheels are generally found on the larger taildraggers.*

## TAILDRAGGER ADVANTAGES

Compared with the tricycle-gear airplane, the conventional-gear airplane has some advantages:

- *More propeller clearance for operating on soft fields and rough fields without damage to the propeller.*
- *The tail wheel is not as susceptible to damage on a rough field compared to a nose wheel.*
- *Better maneuvering at backcountry and remote airstrips. The pilot can execute a turn around in a small space, pivoting on the main gear.*
- *Better takeoff performance from a rough field. The nose wheel plane will drop the nose to the surface increasing surface drag with every dip and bounce.*
- *Better acceleration from a rough field. The pilot can maintain level flight attitude while going over bumps; whereas the tricycle-gear airplane will have an increased angle of attack, becoming momentarily airborne, increasing induced drag.*
- *Faster cruise speed because it lacks the drag of the nose wheel.*
- *Reduced maintenance costs.*

## TAILDRAGGER DISADVANTAGES

- *Weathervane tendency. The taildragger can easily turn into the wind. The tricycle gear has a nosewheel that prevents the weathervane from occurring.*
- *CG location. With the center of gravity located behind the main wheels, physics explains through*

*Newton's Second Law that the object can easily rotate about the point of rotation.*

➤ *Compare the tricycle-gear and conventional-gear planes with a garden cart operating over rough ground in a newly plowed area. The tricycle gear is like pulling the cart. The conventional gear is like pushing the cart. Any irregularity causes the taildragger to veer off track.*

## CONTROLS

Under certain conditions the taildragger requires use of all the controls available. Control effectiveness is based on speed.

As the speed increases during takeoff, the order of gaining control effectiveness is R-E-A, that is, rudder, elevator and finally ailerons.

As the speed decreases the order of losing control effectiveness is A-E-R.

The ailerons, elevator and rudder are considered the controls, but don't leave out the brakes and throttle.

- *Ailerons*
- *Elevator*
- *Rudder*
- *Brakes*
- *Power (throttle)*

*Figure 3.* **A side view of a tricycle-gear airplane reveals the cg (mass of the airplane) is ahead of the main gear.**

## TAXI AERODYNAMICS

The study of physics (that area of science that deals with matter and energy and how they interact) provides some insight that during takeoff the mass of the airplane wants to travel first down the runway.

---

> **Never let an airplane take you somewhere your brain didn't get to five minutes earlier.** — *Anonymous*

---

Tricycle gear airplanes, with the mass ahead of the main wheels, have no problem maintaining a straight track during the takeoff.

Taildragger airplanes, on the other hand, with the mass located behind the main gear, want to swap ends. This is called directional instability and it is a natural occurrence ... in physics, that is.

If the tail wheel is straight as the taildragger starts the taxi, it begins to track straight. The torque of power application, a slight weathercock,

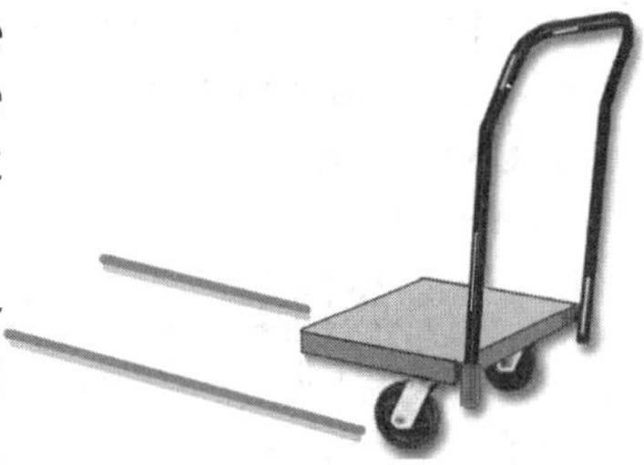

*Figure 4.* **Tricycle-gear airplane comparison. With the mass ahead of the main wheels (this comparison has a person pulling the hand cart), the tricycle-gear airplane tracks straight during takeoff.**

*Figure 5.* **The side view of the taildragger reveals a lot of side area behind the main gear. The cg is behind the main gear.**

*Figure 6.* The mass of the taildragger behind the main wheels wants to go straight (this comparison has a person pushing the hand cart). The wheels may encounter an object to cause them to veer off to the side.

a depression in the surface on one side of the gear, or encountering a small obstruction with one main wheel all have the ability to disrupt the straight track.

Once disturbed the main wheels begin to track off to the side. The mass of the airplane is located behind the main wheels.

Newton's First Law of Motion states that an object in motion tends to stay in motion with the same speed and in the same direction unless acted upon by an outside (un-

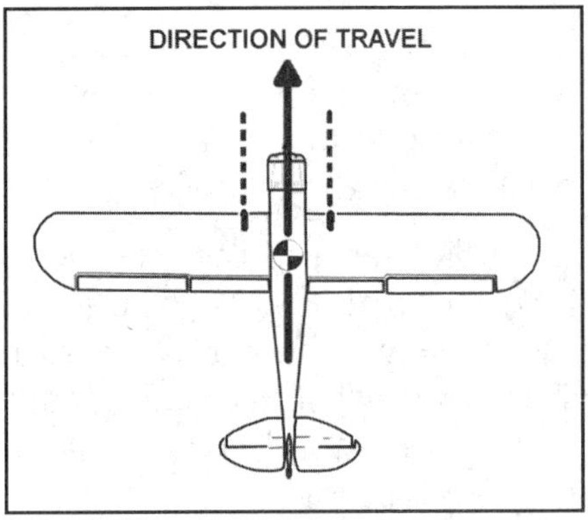


*Figure 7.* If the airplane begins a turn during taxi (or takeoff), the inertia of the mass behind the main wheels tries to continue straight.

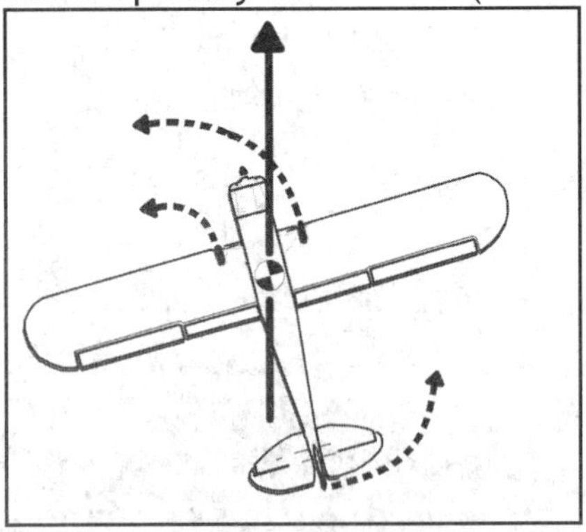

*Figure 8.* During taxi, takeoff or landing any swerve that is not arrested immediately will allow the airplane to veer to the side.

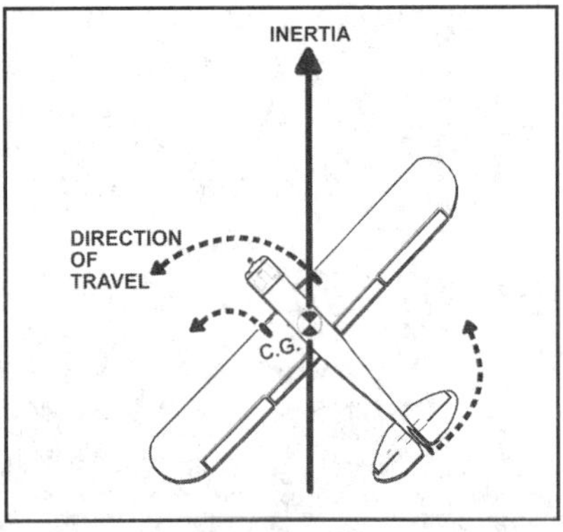


*Figure 9.* Once the airplane veers to the side the inertia behind the main wheels will cause the tail to swing, the start of a ground loop.

balanced) force. The natural tendency for an object to resist a change in its state of motion is described as inertia.

Inertia is the tendency of an object to resist changes in its state of motion, where the "state of motion" is defined by its velocity.

Newton's first law of motion alerts us that when we begin a turn, the mass of the airplane wants to continue along the previous track. Depending on the speed at which the turn is begun, it may be necessary to use opposite rudder or rudder and brake to stop the turn.

Normally the heels of both feet rest on the floor for taxi operations. This allows small quick movements of the rudder pedals. With the feet slid up on the rudder pedals the whole leg is used to make changes. Unless full rudder is required keep your feet on the floor, but be ready to reach the brakes.

Don't find yourself on the backside of the power curve or you will find the airplane will swap ends when you least expect it.

## GROUND OPERATIONS

To establish a habit pattern for the takeoff and landing, taxi is accomplished with the ankles, making small and often fast rudder movements. You will discover that you often get behind the 'power curve' if you use the leg (unless full rudder deflection is required) for rudder deflections.

Taxiing is the first operation a student pilot accomplishes without the assistance of an instructor. Because it is a basic maneuver that is rapidly and easily learned, the directional instability of the taildragger may be forgotten by the occasional pilot who becomes too smug. This complacency often leads to taxiing too fast. This generally happens around the third lesson.

---

**Be wary on the ground.**

---

This indifference is abruptly interrupted and usually abandoned the first time the pilot leans down to look at something on the floor, such as the fuel selector or flaps, or reaches for a dropped pencil. When he looks up the airplane is off the runway or taxiway.

Complacency often leads to inattention. Another example is taxiing behind a large airplane (or even a small turbine-powered aircraft) while its engines are running. Survival of a prop blast or a jet blast encounter helps eliminate this carelessness for years to come.

The pilot has to learn the the taildragger will no more track straight by itself any more than a rope will stay straight as you try to push it along the taxiway.

WARNING Taxiing during gusty wind conditions can cause the taildragger to swap ends in a moment, if the pilot is not watchful.

## Tail Wheel Steering

Most taildraggers have steerable tail wheel assemblies. These allow the airplane to be steered through a range established by the manufacturer, usually about 25 degrees to 30 degrees, although the Cessna 120 and 140 with original manufacturer tail wheel steering, can be turned up to approximately 66 degrees. At an angle greater than these limits, the tail wheel becomes full swiveling. When the turn limit is exceeded—changing from steerable to full swivel—you may hear and feel a "thump."

Once the tail wheel swivels, it is no longer steerable. By moving the airplane forward until the tail wheel is straight, it becomes steerable again.

## Brakes

You may encounter two types of brake-application devices on taildraggers. The 'normal' toe brakes and the Cub-type heel brakes. Beginning pilots go through some strange contortions in learning to use and coordinate the heel brakes.

*Figure 10.* **A glider was pulled in front of a landing Maule. These brakes are quite effective.**

The brake systems of tail wheel aircraft vary with each make and model. The effective range is from nearly nonexistent to extremely effective. Most of the newer airplanes have brakes that will actually stop the airplane.

Always check the brakes when beginning to taxi. If they are not working shut down the engine before taxiing into something.

---

**The first rule of operating a taildragger is: "Fly the darn thing until it's tied down."** — *Howard Ballew*

---

The application of brakes should be with a smooth application of both brakes at the same time. The reason for the subtlety and delicacy is to prevent one brake from engaging first and causing a swerve; and, to prevent the airplane from nosing over. Remember, no sharp application of the brakes.

If only one brake is used, use it cautiously. If the brakes are 'grabby,' have them repaired before flight.

## Differential Braking

Some airplanes are easy to taxi; others, even with steerable tail wheels, may not respond as easily.

Suppose while taxiing on the ramp a 90-degree turn is required. When the turn is made, leading the recovery by depressing opposite rudder before reaching the desired alignment is usually necessary. If a turn is made and the rudder is deflected without stopping the turn, brake application can help in stopping the turn and obtaining the desired alignment.

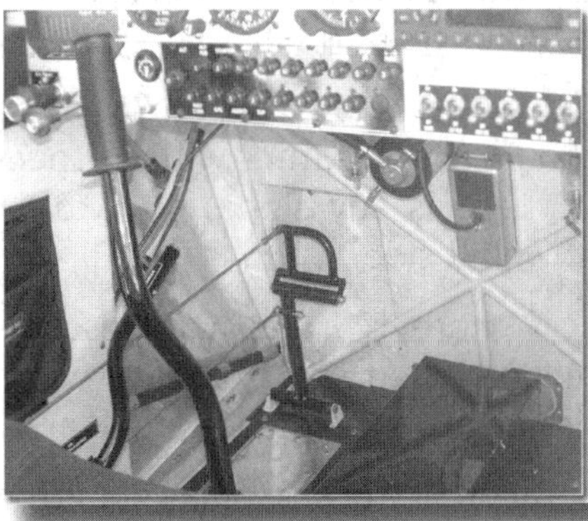

*Figure 11.* **The left-side rudder and brake assembly on the Aviat Husky. Toe brakes are much easier to use than heel brakes.**

It is impossible, because of the many variables, to establish a general rule of thumb for the amount of lead necessary to arrest a turn. Some of the variables include the amount of rudder application/deflection, usage of brakes and brake condition, taxi speed and centrifugal force, wind and taxiway surface. Practice will give the experience necessary to judge the amount of lead required.

A locked-wheel turn can rotate the airplane about the locked wheel. This is improperly used by pilots for parking. A more proper technique is to pump the brake on and off, to save wear and tear on the main tire and assembly. Do not lock the inside wheel during a turn-around procedure.

### Taxi Surface

When taxiing in sand, soft dirt or on soggy ground, once the momentum of the aircraft is established, maintain that momentum even through turns. If the takeoff surface is also soft, do the run-up while taxiing so the airplane does not have to be stopped.

Taxiing on a soft surface will require a great deal of power to start movement of the airplane. Maintain the elevator control in the neutral position to lessen the possibility of rocks being flung by the propeller from striking the elevator surface.

Once the airplane is moving, maintain a constant speed to avoid coming to a stop. If the airplane stops, it may require a very high power setting to begin moving again. This is harmful to the engine and propeller. Debris can be accelerated rearward to damage other parts of the airplane.

Ice or snow-packed taxiways should be treated with respect. If there is any slope, stay to the downhill side.

If there is a berm (landscaped mound used to facilitate water drainage), straddle the peak area.

Again, ice or snow may not afford the opportunity to make a normal run-up while parked and the "before takeoff" checklist will have to be accomplished during the taxi.

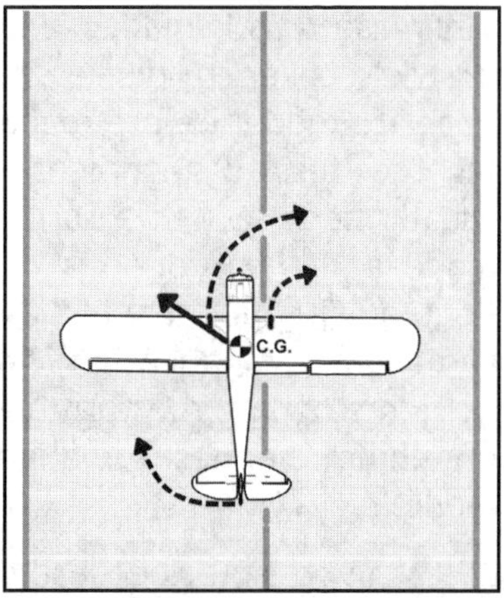


*Figure 12.* **Taxiing along a berm, stay on the peak or on the downhill side.**

## Taxi Speed

Many books state the proper taxi speed is no faster than a fast walk. This is not always feasible. For example, if it is more than one mile to the active runway, it may take up to 20 minutes to reach the departure end of the runway. Taxiing is like driving. You need to use reasonable and prudent judgment. Remember an easy trap to fall into is taxiing too fast. Don't do it.

A safe taxi speed varies under differing conditions of weather and taxi surface. Crosswind and downwind taxiing are done at a slower speed than during a no-wind condition. Taxiing on a soft surface may require more power and speed than taxiing on a hard surface to prevent the surface from stopping the airplane.

## Downwind Turns on the Ground

When taxiing into the wind a considerable amount of rudder pressure may be required to initiate a turn, but once the turn has begun, anticipate the need of neutral or opposite rudder pressure to slow or stop the turn. The amount of rudder pressure will depend upon the taxi surface, the desired radius of turn, and the strength of the wind.

WARNING

Downwind turns, that is, turning from taxiing into the wind to a position away from the wind, require extra caution. The control wheel/stick should be held full aft to place additional weight on the tail wheel, making steering more positive. Be very careful about giving a burst of power to help in the turn, especially when the aircraft has passed the halfway point of the turn where the wind is approaching from the rear or quartering rear. It is possible for the tailwind, combined with the airflow of the propeller blast, to lift the tail to the extent that the aircraft may tip on its nose. If this ever happens you hope it is in an area large enough to allow you to release the brakes and use an additional burst of power to accelerate out of the nose-tipping tendency.

## Upwind Turns on the Ground

On the ground, very little rudder pedal pressure is required to turn into the wind because of the weathercock tendency. Once the turn is established, much more rudder pressure will be required to stop the turn. With experience, you anticipate the need for opposite rudder pressure before the actual need. This allows the experienced pilot to stop the turn without the need of excessive braking.

When taxiing downwind and turning upwind, the weather cock tendency of the airplane will require extra caution. Too much forward speed, combined with the turn into the wind tendency, induces a great deal of centrifugal force on the plane; most are top heavy and may tip over. At the very least, the airplane will experience high levels of side force. These forces are insidious and invisible. The airplane may not break right then, but later on, during normal operations, it may experience a structural failure of the landing gear.

Turning downwind, opposite the weather cock tendency, is difficult. It may require full rudder, differential brake application and some throttle to generate prop blast over the rudder. Improperly executed, the airplane may end up on its nose.

### Visibility From The Cockpit

It seems taildragger airplanes either have excellent visibility or no visibility, with not much compromise between the two extremes. Those airplanes with poor visibility require extra slow-speed taxiing and frequent S-turns to clear the area ahead of the airplane.

### Brakes

The brakes must be tested during the initial movement, before moving more than the length of the airplane. After assuring that brakes are available, their usage should be kept to a minimum.

It is poor technique to use power while adjusting the taxi speed with brakes. Never use the brakes while taxiing when power is applied. Instead, reduce the power to slow the airplane. If the power reduction does not slow the airplane sufficiently, use the brakes. If this scenario is repeated more than once or twice, consider the possibility that the plane is being taxied too fast with too much power application.

Most taildraggers are equipped with either toe brakes or heel brakes. Heel brakes are not especially hard to use, but they do take some getting used to. To begin with, wear a pair of comfortable, flexible shoes. It is impossible to develop a feel for the brakes when wearing combat boots.

## Control Positioning

In a no-wind condition the flight controls are positioned in the full aft (hold the stick against your gut) position with the ailerons neutral.

Although full-aft stick is not technically required in a no-wind condition, it is strongly advocated that the control wheel should be held back "against your gut," whenever the tail wheel is on the ground during landing or taxiing (except with a strong tailwind where the stick is pushed forward). This will be one of the best habit patterns that you can establish.

If the tail-wheel steering is ineffective, you might discover that you haven't pulled back on the control wheel (stick). The additional weight on the tail wheel will make it more responsive. A lack of back pressure causes the tail wheel steering to be less effective, leading to overcorrection to stop any swerves. This can lead to trouble during the takeoff.

To regain control if you get into trouble, close the throttle (in some cases), apply full back pressure on the stick and use rudder with differentail braking to regain control. Use the "rudder usage technique" presented in Chapter 5.

Taxiing with a quartering crosswind requires the ailerons to be moved paralleling the wind. The ailerons help to maintain directional control.

When turning into the wind, the use of opposite aileron movement will help the turn. For example, in making a left turn into the wind, move the aileron control to the right. The left aileron moves down, creating lift and drag, helping the left turn.

Rather than memorizing the control positions required for taxiing in wind, a helpful "thumbs up" technique is presented. It is also included in Chapter 9 – Crosswind Operations for information about control positioning while taxiing in crosswind conditions.

Use extra caution when taxiing in a crosswind from the left. The weather vane tendency is combined with the left-turn forces.

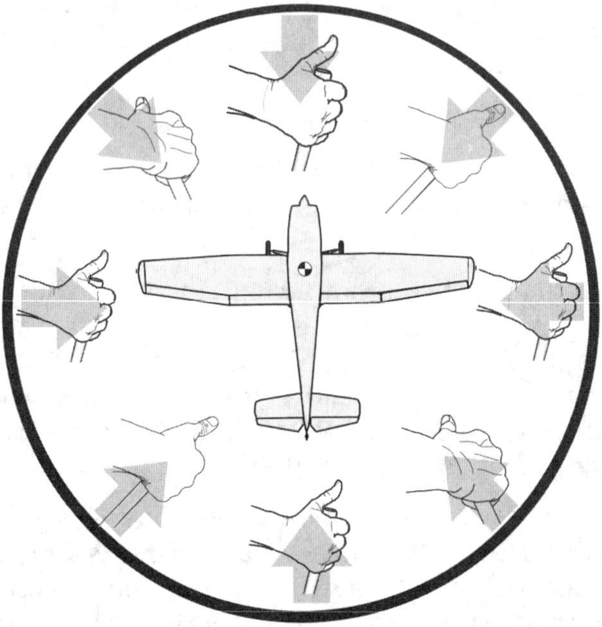

*Figure 13.* **Thumbs-up method of positioning the controls when taxiing. If the wind is from the right front or left rear, parallel the wind with the thumb (move stick to right). If the wind is from the left front or right rear, parallel the wind with the thumb (move stick to left).**

## Taxiways

Engineers at some airports—especially those airports susceptible to a large annual rain fall—have designed the taxiways with a hump or mound (berm) down the center. This berm provides an incline to either side of the centerline that helps in self-removal of the water.

In a no-wind condition, taxiing on the left side of the taxiway makes directional control easier with the rudder because the downslope allows the tail wheel to fight against the natural left-turning tendency of torque and p-factor.

## Turning

Taxiing on a soft surface—such as snow, slush, high grass or mud—will require a good deal of power to start the movement of the airplane. It will be helpful to pump the stick forward and aft to help the aircraft begin movement. Once the airplane is moving, maintain a constant speed to avoid coming to a stop. If the airplane stops, high power will again be required to begin the movement. This is potentially harmful to the engine. Debris picked up by the propeller may not only damage the prop but can be accelerated rearward to damage other parts of the plane.

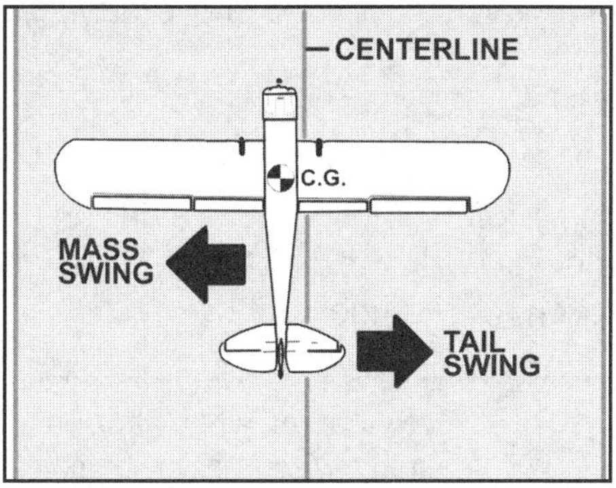


*Figure 14.* **Taxiway with berm. Stay to left side of center line where tail swing tendency due to left-turning forces is somewhat balanced by the mass swing due to the slope of the berm.**

Slow the aircraft before turning to protect the landing gear and associated structure. Centrifugal force and the shift of fuel loads can add tremendous pressure to the gear during a fast turn.

To help a turn in a no-wind condition, once the turn is initiated, use nose-down elevator control (forward pressure) to relieve the weight on the tail wheel.

Locked-wheel turns display poor pilot technique. When differential braking is used to lock the wheel in the desired direction of the turn, a twisting strain is applied to the wheel, axle and strut assembly. Also, the tire is scuffed or worn.

Rather than locking the tire, pump the brake on and off, and

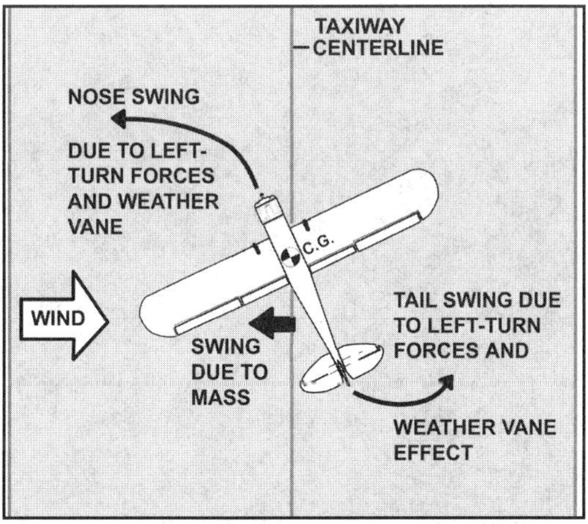


*Figure 15.* **Taxiway with berm and wind from left. Taxi along the steepest area (downhill to the left) where the mass behind the C.G. has a stronger swing effect than the left-turn forces and weather vane action.**

allow some forward movement of the tire to relieve the torque effect.

## Vision

Vision is important in aviation regardless of the type of aircraft flown. With the taildragger's tendency to swerve, vision becomes even more important.

> **No matter what happens, there is always somebody who knew that it would.** – *Anonymous*

When you sit in the airplane on the ramp note the attitude (where the nose and wings are positioned) relative to the horizon. This is the three-point landing attitude.

With tandem seating, use the nose of the airplane for reference. Side-by-side seating requires use of the area in front of you for reference. There may be a line of rivets or some other distinguishing feature on the cowling that can be used until you develop a sight pattern. Using the nose will create a distortion angle that is not helpful during landing.

The airplane can now be safely maneuvered to the run-up area.

*Figure 16.* **The result of having a glider pulled in front of a Maule at the moment of touchdown.**

# CHAPTER 4

# Run-up

## ENGINE OPERATION

Some pilots call an aircraft engine a motor. That's okay, but is not the correct terminology. A motor is an electrical device.

The internal combustion engine is a machine that converts a chemical reaction (the burning the fuel) into heat (expansion to move the pistons) for the rotation of the propeller. One end of a connecting rod (one for each cylinder) is attached to a piston that moves up and down within a cylinder. The other end of the connecting rod is attached to a crankshaft that rotates. The propeller is joined to the end of the crankshaft. The engine, by firing the cylinders, converts straight-line motion of the pistons into rotary motion of the crankshaft to turn the propeller.

### Mixture Control

Most airplanes have a mixture control. It is, or should be, a bright red color indicating that it should be carefully used. Its purpose is for adjustment of the fuel-air mixture.

If the mixture control is not adjusted properly, it can lead to engine damage or failure. For the engine run-up at density altitudes above sea level, the mixture is adjusted to obtain engine smoothness. You are not trying to adjust it for the optimum performance.

## RUN-UP PROCEDURES

The "before takeoff check" is a systematic procedure, before flight, to check the condition of the engine, systems, instruments, radios, and controls. It is often abbreviated by pilots in a rush to get into the air.

Sometimes, without a checklist, important items are forgotten. A written checklist should be used for even simple airplanes to establish a habit pattern that will be transferred to other, more complicated airplanes.

### Park Facing the Wind

When approaching the run-up area park where the propeller is positioned over a part of the ramp or taxiway that is void of debris, small rocks, loose dirt, or sand. If a locked-wheel (poor technique) or partial locked-wheel turn is made, move forward enough to straighten the tail wheel to remove any stress on the assembly. Face the airplane into the wind to aid in engine cooling. Hold the brakes. The stick should be held full aft.

### Control Positioning

When facing into the wind, special control positioning is not required, but the stick should be held full aft while doing the run-up.

### Engine Roughness

If engine roughness occurs during idle or at the run-up power setting, lean the mixture until the engine runs rougher. The heat generated by a lean mixture will help clear the spark plugs. After five or 10 seconds of roughness, enrich until it results in engine smoothness, then do the magneto check.

During the magneto check, the selector is moved from the "both" position to either the "left" or "right" position. The selector is moved back to "both" before checking the other mag. The reason for this is to allow the spark plugs of the mag that was turned off to fire momentarily to clean the electrodes. This provides a more accurate check of the rpm drop. If you notice an rpm drop more than allowed by the manufacturer or if there is engine roughness on one mag (and you have tried to clean the plugs by leaning), have the engine checked by a mechanic before flight.

### Check Flight Controls

Instead of mechanically moving the controls and looking out to see if the ailerons move up and down, use a trick to detect not only movement, but control continuity. Although it is rare, control systems have been connected backwards during maintenance.

With the saying "up right," you can move the stick to the right and the right aileron should move "up."

Checking control movement of the rudder, the ailerons and the elevator throughout their full range is as important as checking the continuity of controls. Use a "box" as a mental pattern and move the controls to each corner, perhaps starting in the left front corner of your "box," then moving to the right front corner, right rear corner, left rear corner and forward to the left front corner. This insures that nothing will cause binding within the control range and that nothing situated in the cockpit interferes with the control movement.

### Adjusting Mixture for Takeoff

Operations at high mountain airports and unimproved strips may have colored my impression that static run-ups at power settings above the recommended mag check setting are potentially harmful to the engine and propeller. Therefore, my preference is to adjust the mixture during the takeoff run, rather than while sitting at the end of the runway.

To adjust the mixture, use an approximate setting that allows engine smoothness. After application of takeoff power, ease the mixture control out until the start of engine roughness, then ease it in until the engine runs smoothly. My guess is that during the first few times you try this, the mixture will be adjusted too much and the engine will momentarily stop. Just ease in the mixture until the engine runs smoothly.

---

**An optimist, after all, is often as wrong as the pessimist, but he has a lot more fun.** – *Anonymous*

---

Using this technique, your attention may be directed outside the airplane without needing to have to look at the tachometer, exhaust gas temperature (EGT) gauge, or at anything else inside the cockpit. Fine tuning of the mixture can be accomplished after climbing 500 or 1,000 feet above the ground.

Some pilots adjust the mixture at a power setting of approximately 1,800 rpm. This is not an accurate manner for adjusting the mixture. A power enrichener valve is incorporated in many carburetor designs to provide extra fuel for engine cooling at full throttle operation. It will negate the partial power mixture adjustment. Even without the power enrichener, changes in air flow and fuel flow with an open carburetor butterfly valve will make partial power mixture adjustment invalid.

## SAMPLE CHECKLIST

Appendix A contains a sample checklist for a converted Cessna 170B. Because of the AVCON conversion (Lycoming O-360 engine, fuel pump and constant-speed propeller), some of the items would not pertain to an airplane with a fixed-pitch propeller. It should, however, be helpful in developing a checklist for your particular airplane.

When a pilot is taught to use a checklist while learning to fly, this habit pattern carries over to other airplanes, even if they are not complex. The use of a checklist is important for safety.

When a pilot learns to fly with memorization, rather than using a written checklist, it is very difficult to convert him to the use of a checklist on a regular basis. When these pilots transition to a complex aircraft with retractable landing gear, they occasionally get into trouble.

If you find yourself in the latter category, at least use a method for your memorized checklist. For takeoff something like the mnemonic CIFFTRS, will insure the minimum items are checked. The word stands for *Controls, Instruments, Fuel, Flaps, Trim, Run up,* and *Seat belts.* For landing use GUMPS, *Gas, Undercarriage, Mixture, Prop,* and *Seat belts.*

*Figure 1.* **I once sold a Super Cub to a fellow who did not want a tail wheel checkout. The next time I saw the Cub it had dual-tandem landing gear installed.**

# CHAPTER 5

# Take Off

## AERODYNAMIC FORCES

To eliminate duplication of material about aerodynamic forces, please review the following subjects of Chapter 1—Aerodynamic Forces:

- ✔ *Takeoff moment*
- ✔ *P-factor*
- ✔ *Gyroscopic precession*
- ✔ *Tricycle gear*
- ✔ *Taildragger comparison*
- ✔ *Centrifugal force.*

## RUDDER-USAGE TECHNIQUE

There is no mystery involved in the takeoff of a coventional-gear airplane. There are, however, certain elements of flight that may require the acquisition of new and more efficient piloting skills and techniques.

Directional control during takeoff in a conventional-gear airplane, that is, maintaining a straight track during the takeoff or landing roll, is a problem caused by *torque reaction* and *takeoff moment* (the increased drag on the left tire due to torque), *p-factor* (asymmetric loading of the propeller), *gyroscopic precession* of the propeller and the *corkscrewing effect* of the propeller slipstream (spiral effect or spiral slipstream effect).

The directional control problems inexperienced pilots encounter during takeoff and landing most often happen because of the delay in making a correction when a turn or swerve develops or once the correction is made, waiting too long before neutralizing the rudder input causing a swerve in the opposite direction.

Don't wait to take action until the airplane's nose has moved 10 degrees or more from its lateral alignment. It is important to counteract any turn tendency when the nose first begins to move.

Overcontrolling with the rudders can cause as much of a problem as a delay or inaction. Overcontrolling most often occurs because the pilot looks too close in front of the airplane. This causes any movement of the nose to be exagerrated with the resulting overcontrol. It will be easier to maintain the nose straight if you look down the runway toward the end, using peripheral vision to detect nose movement.

As you begin a takeoff the airplane will want to move left due to torque and the takeoff moment, p-factor, gyroscopic precession (only when the tail is raised) and the spiral slipstream. When the tail is raised, the gyro precession will turn the airplane farther left. The faster the tail is raised, the greater the gyro precession effect, teaching us to raise the tail slowly and smoothly. With experience you will develop the ability to recognize these predictable turns or swerves before they develop and you are able to apply rudder correction at or just prior to its need.

The novice pilot, not recognizing the nose movement, may wait too long before trying to rectify a turn or swerve. More and more rudder pressure is required to cancel the swerve. If the pilot uses prolonged application of the rudder, in the direction opposite the turn, the overabundance of rudder causes a swerve of greater magnitude in the opposite direction ... setting the pilot up for a ground loop.

## The Problem

The term 'taildragger' is actually a misnomer. The name has been carried forward from the early days of aviation when airplanes had a tail skid instead of a tail wheel.

Taildraggers are different from tricycle-gear airplanes. You first notice this difference when taxiing and making turns. The initial turn from the parking space may cause some surprise when the turn continues until (or even after) using opposite rudder to straighten the path. At such a slow speed some brake application may also be necessary.

Most of the techniques described here have evolved over the years from experimentation to find the best procedure. An explanation of those procedures that do not work is provided so you do not have to repeat the often bitter experience of many pilots.

## The Solution

My greatest success in teaching pilots rudder usage in the taildragger comes from a technique of pushing on both rudders at the same time. Use equal pressure against each rudder pedal (push on both rudders simultaneously), and then move them back and forth, depressing each rudder about one or two inches. The two-inch depression will be referred to as the 'normal deflection.' Obviously when one rudder is deflected there must be a partial relaxation on the opposite rudder pedal, but, the pressure is still kept on both rudders.

Align the airplane along the centerline of the runway for take-off. Suppose you need to push the left rudder a bit for alignment. While the left rudder is pushed, partial pressure is maintained on the right rudder pedal. Prior to the desired alignment, lead the turn to straighten the tail wheel with opposite rudder. Now the airplane is aligned with the runway and ready for the application of power.

As the power is applied for takeoff, the rudders are moved back and forth, at a rate of about one depression per second. This is a rapid movement of the rudders. This means that if the right rudder is depressed for normal deflection—two inches—the left rudder is immediately depressed four inches. This is two inches of left rudder to reach neutral, and two more inches for the normal deflection.

This technique recognizes that flexing the ankle is the preferred manner of rudder control, making small and fast pokes on the rudder pedals. Often a pilot will use leg movement rather than ankle movement. Unless full rudder application is needed, use your ankles. You can't move your leg as fast as your ankle and you can get behind the airplane.

When a turn or slight swerve is recognized, the rudder movement opposite this turn or swerve must be greater than the two-inch normal deflection—perhaps a three- or four-inch depression of the rudder opposite the turn or swerve.

Nevertheless, the technique requires the pilot to go back to the other rudder and depress it the normal two inches from the neutral position. If the turn has not been arrested, the greater rudder deflection opposite the turn is again effected with immediate return to the other rudder. This procedure cancels the tendency of holding the corrective rudder until the swerve has been corrected and the airplane darts off in the opposite direction.

If the swerve continues, jab the brake along with the rudder. If brake and rudder movement does not straighten the airplane, do not hesitate to use full rudder pressure (to the stops) and some brake action along with the rudder deflection. But don't forget to go back to the normal deflection to prevent a swerve in the opposite direction.

This method of making a correction and moving the rudders back to normal deflection prevents over correction where the pilot has to deal with a large swerve that develops in one direction, then switches to a larger swerve in the other direction.

After three or four flight lessons to develop a feel for the airplane, try a takeoff without the constant rudder movement, using the rudders as required. Remember that once the correction is made, the rudders are moved back to neutral.

As soon as the airplane begins to move in either direction, use opposite rudder as necessary. The amount of rudder is dependent on speed. At slow speeds the rudder deflection (application) may be one half to three quarters of the total that is available. At fast speeds the rudder may be pushed about one quarter or less of that available.

Regardless of the amount of rudder used, it is important to return to a neutral position before the airplane over-reacts and starts a swerve in the direction opposite the original swerve.

## RUNWAY ALIGNMENT

Normal takeoffs are made from the centerline of the runway. When taking off with a crosswind, try to align the airplane into the wind. If this requires alignment along the edge of the runway pointing toward a place down the runway toward the other edge, do it. *See: Chapter 9, Crosswind Operations*

Some aircraft designs prevent forward visibility, in which case peripheral vision is used to maintain runway alignment.

Rather than move your head from side-to-side in airplanes with restrictions to forward vision, scan the area with eye movement. This allows faster detection of turns or swerves.

## POWER APPLICATION

During takeoff a smooth application of power is necessary to prevent a large swerve to the left caused by torque. Applying power smoothly will also protect the engine from internal damage. If the airplane is moving forward over the ground before full power is achieved it will also minimize the chance that rocks or debris will be funneled to damage the prop.

Begin with the control stick positioned full aft. The heels are normally rested on the floor. The novice pilot may want to keep his feet close to the brakes.

Smoothly apply full power (observe the power limitations on turbocharged/supercharged engines). As the speed increases during the takeoff roll, relax the back pressure, eventually moving the stick forward of the neutral position to raise the tail (properly trimmed, the tail will come up by itself as the airplane accelerates). A rapid movement on the control stick from full aft to forward of neutral will cause the airplane to swerve left due mainly to the gyroscopic precession of the propeller and to a lesser extend due to the torque and the p-factor. P-factor is reduced, but still present.

Anticipate the need for right rudder input with the original application of power and again as the tail comes up.

Knowing the order of control effectiveness during the takeoff helps in maintaining control of the airplane. The order is R-E-A. The rudder is the first control to become effective, followed by the elevator, then the ailerons.

Due to the propeller blast during the initial takeoff roll the rudder is somewhat effective; however, the tail wheel must be held firmly on the ground to provide positive steering until the rudder becomes effective enough to provide air steering.

As the speed increases the rudder becomes more and more effective. Smaller rudder corrections (rudder deflections) are needed to control the aircraft's ground track at the faster speeds of takeoff and landing.

Because the elevator becomes effective before the ailerons, do not try to remedy a poor directional control problem by pushing the stick forward prematurely to raise the tail and gain better visibility. Rushing the takeoff removes the positive tail wheel steering before effective rudder steering is available.

Although the power application is made smoothly, it should also be made promptly. Two or three seconds from the idle position to the full power position will safeguard the engine from damage.

## SWERVING

If a pilot fails to recognize the beginning of a swerve in time to make a normal correction with rudder usage, the application of the brake opposite the swerve may also be needed to straighten the plane.

When a swerve starts, some pilots have frantically reduced the power to idle, determined to straighten and stop the plane. This may or may not be the best solution depending upon the situation.

If the aircraft speed is slower than that required for the rudder to be effective, reduce the power to idle and pull back on the control wheel to place more weight on the tail wheel for more effective steering. Use the rudder, brakes and ailerons as required to reacquire control of the ground track.

When the airplane has accelerated to a speed where the rudder is effective in controlling the airplane and a swerve develops, it is usually best to leave the power on. The airplane is more controllable with power on because of the blast of air over the rudder. If the power is reduced, the rudder becomes less effective in controlling the swerve.

*Figure 1.* **The maximum lift for any airfoil design occurs with full aileron deflection for an optimized lift-to-drag ratio.**

If the speed is fast enough for the rudder to be effective, it is probably too fast to pull back on the control wheel to place weight on the tail for better steering. It is undesirable to pull back and rotate the airplane without sufficient lift to fly. In a crosswind this will cause skipping and will compound the problem.

*Figure 2.* **Extend the flaps to match (parallel) the aileron deflection. This is the maximum lift for a normally aspirated engine.**

If the aircraft speed is close to flying speed, the application of half of the flaps may allow you to fly out of the directional control problem. If flaps are used, it is important to immediately lower the nose to level-flight attitude or slightly above level-flight attitude and accelerate in ground effect before climbing or retracting the flaps.

*Figure 3.* **Extend the flaps to parallel the aileron deflection for Cessna aircraft.**

## Flap Settings

The purpose of flaps is to allow the airplane to make a steeper approach angle for landing without increasing the airspeed. They do so by increasing the camber of the wing and thus increasing the lift. Because lift and drag are directly proportional, the increased lift also increases drag. With reduced power, the descent angle is greater without adding extra airspeed.

A side benefit of flaps is that they lower the stall speed and allow for a slower approach and touchdown speed. This also means that the use of flaps during takeoff will result in a shorter takeoff ground roll.

Follow the manufacturer's recommendation on how much flap to use. Older airplanes may have no documentation or no recommendation on the use of flaps. In this case, move the control wheel or stick to either side for full aileron deflection. Match the flap deflection as closely as possible to the aileron deflection. This will provide maximum lift for the particular airfoil design of the airplane.

## TECHNIQUE TO PREVENT GROUND LOOPING

The best technique for the new taildragger pilot to prevent a ground loop is the rudder-usage technique explained at the beginning of this chapter. Use the rudder and brakes as necessary to stop the swerve. The important thing is to neutralize the rudder immediately after the swerve is arrested. Otherwise the plane may swerve in the opposite direction. This is the reason for using the technique of moving the rudder pedals back and forth, with more deflection in the direction that is opposite the swerve.

Do not be hesitant about using the brakes when a swerve develops. The wear-and-tear on the brakes and tires may be less than the tear-and-wear on the airplane if a ground loop occurs.

Using everything available is important. Often the ailerons are forgotten as an aid in controlling the airplane. If the speed is fast enough for the ailerons to be effective in rolling the airplane, roll the wings level or opposite the swerve. This is instinctive, so don't think about it, just use them.

If the speed is too slow for the ailerons to be effective in rolling the plane, move the control wheel toward the swerve. This is something you have to preplan and it requires that you think about it beforehand because it is not instinctive. The deflected aileron, although not effective enough to produce the lift required to roll the airplane, does produce some lift. Being that lift and drag are directly proportional, the increased drag from the deflected aileron at slow speed will assist in straightening the ground path. Although the drag is very little, everything you can do will help.

### Put It All Together

It takes integration and coordination of all the flight controls to operate a taildragger on the ground. You might start by trying the "rudder-usage" technique.

The airplane becomes airborne smoothly and with positive control during the normal takeoff. This is done by rotating the airplane to establish a flying attitude. Usually the takeoff is not consciously performed by pulling back on the control wheel until the airplane is forced off the ground. Rather, the airplane is allowed to fly off the runway when it is ready to fly.

General aviation taildraggers, depending on excess power available, fly at the best rate-of-climb airspeed, $V_Y$, at about an 8.5-degree nose-up pitch attitude. The airspeed at liftoff is less than the best rate-of-climb airspeed due to ground effect and a pitch attitude of somewhere between 3-degrees and 7.5-degrees nose up depending upon the pilot's preference.

This technique compensates for ground effect. It is possible to fly an airplane off the runway, using ground effect, at too slow an airspeed to sustain flight out of ground effect. The airplane quits flying about two or three wing spans above the surface and settles or falls back to the runway with undesired effects.

By maintaining the takeoff attitude—the position of the nose in relation to the natural horizon—the airplane will continue to fly once clear of ground effect.

## FACTORS REQUIRING CONSIDERATION FOR TAKEOFF

### Aircraft performance

An airplane performs according to the altitude where it thinks it is flying, or the density altitude. Density altitude is an expression of the thickness of the air the airplane flies through. For example, it is possible to be operating from an airport with a physical elevation of 6,000 feet msl that has a density altitude of 10,000 feet. In this instance the aircraft performs as though it were at 10,000 feet in the standard atmosphere. (See: Density Altitude)

Look at the performance charts provided by the airplane manufacturer in the Owner's Manual or Pilot Operating Handbook with suspicion. The takeoff distances quoted are for a new engine and may have been under ideal conditions and the airplane was probably flown by a professional pilot.

It is often a good idea to have a back up plan in most airplane operations. One that works well for the takeoff distance is to use the 50/70 Rule for Takeoff Rule of Thumb (page 5-30) that states if 71 (70.8) percent of the takeoff speed is attained in the first half of the runway, the airplane has sufficient runway to takeoff in the remaining half of the runway. This rule of thumb only guarantees the takeoff. It does not take into consideration any obstacles at the end of the runway or during the climb out.

### Rule of Thumb - Determine Density Altitude

*For each 10°F above (or below) standard temperature at an airport's elevation, add (or subtract) 600 feet to (from) the field's elevation.*

The FAA definition of density altitude is *the pressure altitude corrected for nonstandard temperature variation.* To find the density altitude, use a DENALT performance computer, electronic flight calculator, graph, or the rule of thumb.

---

**An airplane will probably fly a little bit over gross weight, but it sure won't fly without fuel.**
— *Anonymous*

---

The "Determine Density Altitude" rule of thumb will rarely result in an error of more than 200-300 feet. For each 10°F above (or below) standard temperature at an airport's elevation, add (or subtract) 600 feet to (from) the field's elevation. This value will be the altitude that the airplane thinks it is flying.

The airport's standard temperature can be determined by converting the field elevation and multiplying by the standard lapse rate.

For example, Centennial airport, Englewood, Colo. has an elevation of 5,883 feet. Dividing by 1,000, the elevation becomes

5.883. Multiply this by 3½ degrees, the standard lapse rate, to determine the temperature at Centennial is 20.6 degrees cooler than the standard sea level temperature of 59°F. Subtract (59 - 20.6) and you find the standard temperature at Centennial is 38.4°F.

If the temperature on a given day at Centennial is 70°F, what is the density altitude? Using the rule of thumb, subtract the airport's standard temperature (70°—38.4°) from the existing temperature, for 31.6. Divide this by 10 (to determine each 10°F increment) for 3.16. Multiply this by 600 feet per 10°F for 1,896. Add 1,896 to the field elevation, 5,883, for a density altitude of 7,779 feet. You will note this rule does not use the pressure altitude, yet it is surprisingly accurate.

Using an electronic flight calculator and an altimeter setting of 29.92, the density altitude is 7,910 feet (within 131 feet of the rule of thumb). With an altimeter setting of 30.10, the density altitude is 7,691 feet (within 88 feet of the rule of thumb). With an altimeter setting of 29.80, the density altitude is 8,056 feet (within 277 feet of the rule of thumb).

Other rules of thumb that may be useful in determining aircraft performance include the following.

### Rule of Thumb - Density Altitude Takeoff Distance - Fixed-pitch Propeller

*To the standard sea level takeoff distance, add 12 percent for each 1,000-foot density altitude up to 8,000 feet. Add an additional 20 percent for each additional 1,000-foot density altitude above 8,000 feet.*

### Rule of Thumb - Density Altitude Takeoff Distance - Constant-speed Propeller

To the standard sea level takeoff distance, add 10 percent for each 1,000-foot density altitude up to 8,000 feet. Add 15 percent for each additional 1,000-foot density altitude above 8,000 feet.

These rules of thumb do not provide exact performance information. Working a few examples is important to compare the rule of thumb numbers with the owner's manual numbers to determine if the rules are applicable to your airplane.

### Rule of Thumb: Density Altitude Rate of Climb - Fixed-pitch Propeller

*Reduce the sea level rate of climb 7 percent for each 1,000-foot density altitude up to 8,500 feet and 8 percent for each 1,000 feet above 8,500-foot density altitude.*

### Rule of Thumb: Density Altitude Rate of Climb - Constant-speed Propeller

*Reduce the sea level rate of climb 6 percent for each 1,000-foot density altitude up to 8,500 feet and 8 percent for each 1,000 feet above 8,500-foot density altitude.*

## Airspeed

Use the same indicated airspeed as that used at sea level to rotate for takeoff despite the density altitude or physical altitude at an airport. At high-elevation airports a greater true airspeed is required to compensate for the thinner air; however, the same thin air that decreases the lift of the wings, the engine performance, and the thrust output of the propeller also affects the airspeed indicator.

An airplane accelerating for takeoff from a sea-level airport under standard conditions (assuming no instrument error) will register the true airspeed of the airplane. The indicated airspeed and true airspeed are the same. When the airplane operates at altitudes above sea level, the indicated airspeed is less than the true airspeed. The actual true airspeed is approximately two percent faster than the indicated airspeed for each one-thousand feet above sea level.

The airspeed indicator operates like a roll-up party whistle. The harder one blows on this whistle, the more it unwinds. The same thing happens with the airspeed indicator. As more air enters the pitot tube, the bourdon tube "unrolls" and indicates a higher airspeed.

At high altitude airports, more thin air must rush through the pitot tube to register the same indicated airspeed as at sea level. The same indicated airspeed at a high altitude airport equates to an increased true airspeed of the airplane. By using the same indicated airspeed, there is an automatic compensation for a density altitude.

## Gross Weight

The takeoff distance varies with the square of the gross weight. Under certain density altitude conditions at high-elevation fields, it may be necessary to restrict the load to obtain a margin of safety for the takeoff. A 10 percent increase in takeoff gross weight will cause:

- *a 5 percent increase in the speed necessary for takeoff;*
- *at least a 9 percent decrease in acceleration; and,*
- *at least a 21 percent increase in takeoff distance.*

Conversely, a 10-percent decrease in takeoff gross weight will cause a decrease of the same percentages.

## Rule of Thumb - The Takeoff Distance Varies as the Square of the Gross Weight

Suppose you are operating a Cessna 170B at 1,500-pounds gross weight. The maximum allowable gross weight is 1,900 pounds.

To compute the effect of the reduction of weight on takeoff distance, obtain the percent of gross weight. Divide the actual weight by the maximum allowable gross weight. This is 1,500 divided by 1,900, or 0.79. Squaring this number, 0.79 x 0.79, equals 0.62. The takeoff distance is equal to 0.62 multiplied by the normal takeoff distance for the density altitude of the airplane.

## Wind

The presence of a headwind will assist in the takeoff and landing by shortening the distance required. For an approximation of the effect of the wind, use the following rule of thumb.

### Rule of Thumb - Head Wind Reduces Takeoff Distance

*A head wind will reduce the normal takeoff distance for any particular density altitude equal to 90 percent of the takeoff distance minus the value of the ratio of the head wind component divided by the rotation speed.*

When you do not want to do mathematical computations, the following is a rough estimate of the effect of the headwind in reducing the takeoff distance.

- ✔ *10 knots = 27%*
- ✔ *15 knots = 39%*
- ✔ *20 knots = 52.5%*
- ✔ *25 knots = 65%*

When operating with a crosswind, the wind is divided into two vector forces (except a direct headwind or a direct crosswind). One component is the crosswind. The other component is the headwind.

| Angle | Sine | Cosine |
|---|---|---|
| 0 | 0.000000 | 1.000000 |
| 10 | 0.173648 | 0.984808 |
| 20 | 0.342020 | 0.939693 |
| 30 | 0.500000 | 0.866025 |
| 40 | 0.642788 | 0.766044 |
| 50 | 0.766044 | 0.642788 |
| 60 | 0.866025 | 0.500000 |
| 70 | 0.939693 | 0.342020 |
| 80 | 0.984808 | 0.173648 |
| 90 | 1.000000 | 0.000000 |

*Figure 4.* **Excerpt from the SINE and COSINE tables.**

To determine the headwind/crosswind component, use a flight calculator or computer. A SIN/COSINE table can be used for a mathematical computation. The HEADWIND (or tailwind) component is equal to the COSINE of the wind angle (the angle between the runway heading and the wind direction) multiplied by the wind velocity. The CROSSWIND component is equal to the SINE of the wind angle multiplied by the wind velocity.

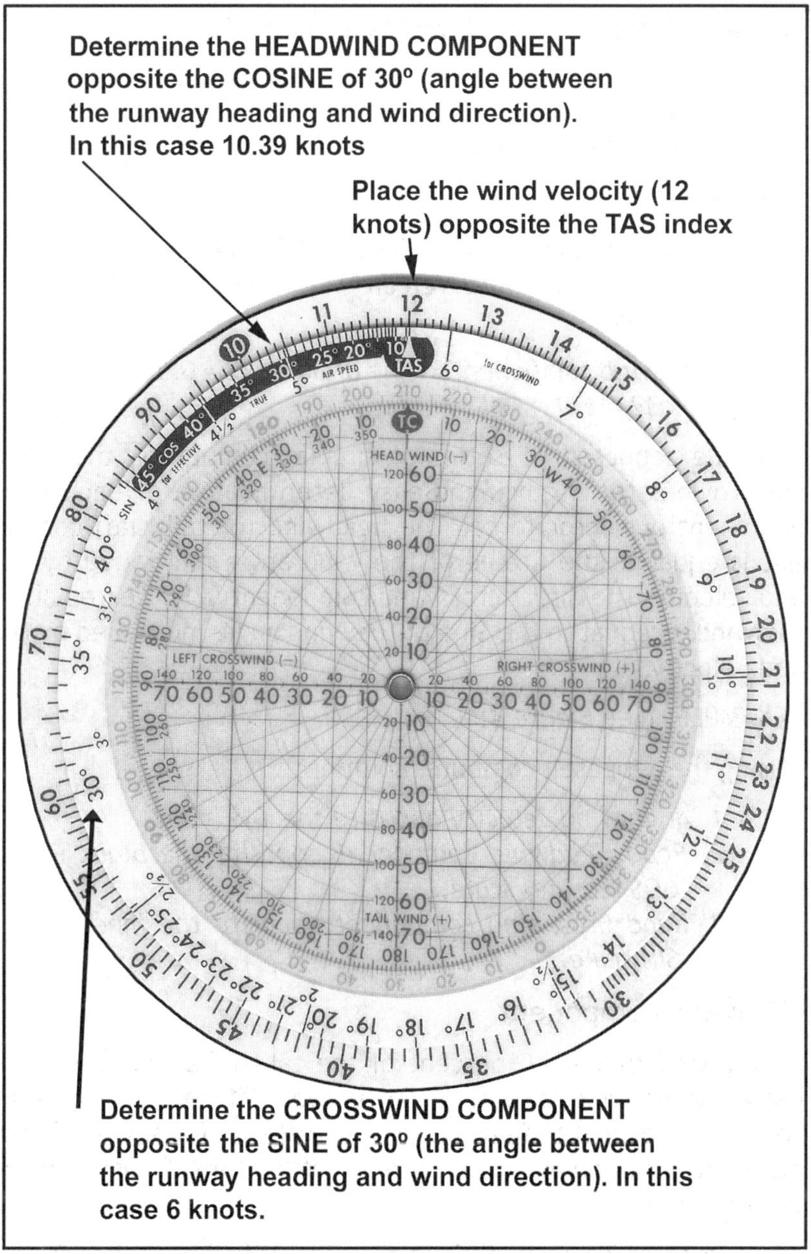


*Figure 4.* Jeppesen CR computer

EXAMPLE:
Using the mathematical Sine/Cosine method:

Conditions:

*Runway: 18*
*Wind: 210º at 12 knots*

Solution:

*Headwind component = Cosine of 30º (wind –runway) x 12 (wind velocity) = 0.866025 x 12 = 10.392*

*Crosswind component = Sine of 30º x 12 = 0.500000 x 12 = 6.0*

For the Jeppesen CR-series of circular computers, the backside provides for a quick-and-easy determination of each wind component or vector force. The headwind (or tailwind) component is equal to the Cosine of the wind angle multiplied by the wind velocity. The direct crosswind component (a 90-degree angle crosswind) is equal to the Sine of the wind angle multiplied by the wind velocity.

EXAMPLE: (See *Figure 4*)

Using the Jeppesen computer solution:

1. *Align wind velocity opposite the TAS index.*
2. *Read Headwind component opposite the black COS (cosine) scale.*
3. *Read Crosswind component opposite the inner SIN scale.*

### Tailwind component

A tailwind that is 10 percent of the takeoff airspeed will increase the no-wind takeoff distance about 21 percent. By using the aircraft manual to find the takeoff distance for any particular density altitude, an accurate estimate of the takeoff distance can be determined.

### Rule of Thumb: Effect of a Tail Wind on Takeoff Distance

*The tailwind takeoff distance is equal to 110 percent of the takeoff distance plus the value of the product of the takeoff distance multiplied by the tailwind component divided by the rotation speed.*

EXAMPLE:

A Cessna 170 (assume the C170 rotates at 50 knots) makes a takeoff with a 10-knot tailwind component from an airport at 5,000-feet elevation and 20-degrees Celsius temperature. The normal ground run for takeoff is 1,430 feet. What effect does the 10-knot tailwind have?

Add 10 percent of the takeoff distance = 1,430 + 143 = 1,573. Add the value of the tailwind component (10 knots) divided by the rotation speed of 50 knots = 0.2. Then 1,573 times 0.2 = 314.6. Add 1,573 and 314.6 to obtain 1,887.6. The 10-knot tailwind increases the takeoff distance by 457.6 feet.

## Gusty Wind Takeoff

For takeoff during gusty wind conditions, keep the airplane on the ground until obtaining a speed of $V_R$ (rotation speed) plus one-half the gust factor.

## Runway Surface and Condition

The takeoff figures in an airplane flight manual have been developed for takeoff on a hard-surfaced runway. Long grass, sand, mud, snow, slush, or gravel will adversely affect the takeoff distance. Frictional effects increase the takeoff distance when operating on other than paved runways.

### Rule of Thumb: Takeoff Distance for Various Surfaces

Increase the takeoff distance due to frictional effect by the following amounts:

- ✔ *Firm turf - add 7%*
- ✔ *Rough, rocky, or short grass (up to 4 inches) - add 10%*
- ✔ *Long grass (more than 4 inches) - add 20% to 30%*
- ✔ *Soft field - add 23% to 75%*
- ✔ *Mud or snow - add 50% or more*

For combinations of various surfaces, add the percentage of each variable. For example, if the surface is soft from recent rain and grass covered, to the normal takeoff distance add 10 percent for the grass, and 23 percent for the soft surface.

These computations provide "ball park" figures. Because they are only an approximation of the performance you may expect, inexperienced pilots should add a 25- to a 50-percent safety factor until they have gained experience in judging the takeoff surfaces. Or, use the 50/70 rule, obtaining 70 percent of the speed necessary to rotate at the half-way point of the runway.

### Gradient (Sloped Runway)

Not all airports are "flat." A slight incline of only a few degrees can affect the takeoff. Generally, it is best to takeoff downhill from these airports. A one-percent downslope is equivalent to about 10 percent more runway in terms of takeoff performance. If the wind or obstruction creates a need to takeoff upslope, the additional drag and rolling friction caused by the one-percent upslope can result in a two-percent to four-percent increase in the takeoff distance.

#### Rule Of Thumb - Upslope Runway Takeoff

*To the normal takeoff distance for any particular density altitude, add 10 percent more takeoff distance for each degree of upward slope in the runway.*

#### Rule Of Thumb - Downslope Runway Takeoff

*To the normal takeoff distance for any particular density altitude, subtract five percent of the takeoff distance for each degree of downward slope in the runway.*

The following two lists are more accurate than the general rule of thumb for working with downslope or upslope takeoff distances.

### Downslope Reduces Takeoff Distance

- *2% downslope reduces takeoff distance 11%*
- *4% downslope reduces takeoff distance 19.5%*

- *6% downslope reduces takeoff distance 28.5%*

### Upslope Increases Takeoff Distance

- 1% upslope increases takeoff distance 7.5%
- 2% upslope increases takeoff distance 14%
- 4% upslope increases takeoff distance 25.5%
- 6% upslope increases takeoff distance 39.5%

### Humidity

If the temperature and pressure are constant, air density varies inversely with humidity. If the humidity increases, the air density decreases. Decreased air density reduces the performance of the airplane.

### Aircraft and Engine Condition

An older airplane with a worn engine cannot be expected to perform according to the performance charts in the pilot's operating handbook or owner's manual.

### Experience and Performance

The more flight experience one has, the greater the performance that can be obtained from the airplane. However, experienced pilots must avoid complacency or a know-it-all attitude.

### Effect of Local Terrain

Terrain shapes that lend to the presence of strong downdrafts and wind shear once the aircraft becomes airborne on takeoff are especially critical.

Wind circulation patterns resulting from the effect of local terrain may reduce aircraft performance sufficiently to prevent the aircraft from out-climbing the rising terrain. The *Aeronautical Information Manual, Flight Guide* and other publications provide information for the correct departure procedure for airports posing this problem.

For example, *Flight Guide* (Airguide Publications, Long Beach, Calif.) provides the following warning for Lander, Wyo.:

> **DEP: Rwy 21-Rgt tfc ok in hi wind to avd mtn in severe dwndrfts.**

## GROUND EFFECT

Most pilots have encountered the term "ground effect." As a term it is frequently tossed about in hangar sessions. Although the term is widely used, its effect on airplane performance is dimly understood by some pilots.

Accidents attributed to ground effect have a clearly visible pattern. The field is short and rough, or it is covered with snow, mud, or tall grass. After a sluggish run, the pilot forces the airplane off the ground, barely clears the fence at the end of the runway, then seemingly clear of the ground, falls disastrously back to earth, often with dreadful results.

---

> **Mother nature is a great teacher, but she gives the test first ... and the lesson afterward.** – *Anonymous*

---

### Drag

The total drag acting on an airplane is divided into two components, parasite drag and induced drag. The two are always present, though in varying degrees.

Parasite drag is caused by skin friction and the turbulence of the air flowing past the various parts of the airplane.

Induced drag is the result of the wing's work of sustaining the airplane in flight. The wing lifts the airplane simply by accelerating a mass of air downward. It is perfectly true that reduced pressure on top of an airfoil is essential to lift (Bernoulli's principle), but that is but one thing that contributes to the overall effect of pushing an air mass downward.

The more downwash there is, the harder the wing is pushing the mass of air downward. At high angles of attack, induced drag is high. Since this corresponds to lower airspeeds in actual flight, it can be said that induced drag predominates at low speed and parasite drag is greatest at high speeds.

### Ground Effect

Ground effect has been incorrectly explained by some educators merely as a "cushion of air" between the wing and the ground.

Ground effect works on wing downwash and wingtip vortices. It reduces the angle of downwash and diminishes the effect of wingtip vortices. This results in a reduction of the airplane's induced drag when very near the ground.

Some explanations of lift state the relative wind is parallel to the longitudinal axis of the airplane. It is not. Rather, there is an upwash ahead of the wing and a downwash behind the wing. When the wing is flown close to the ground, the upwash and downwash cannot fully develop. Because of this, the wing is able to produce sufficient lift for flight at an angle of attack that is less than the angle of attack required at a higher altitude. The induced drag has been reduced. It is important to know the altitude where ground effect disappears in order to take advantage of its benefits and to avoid an unintentional stall when flying at the slower speed the ground effect allows.

NOTE: At the moment of liftoff there is about a 48-percent reduction in the induced drag when compared to flight at altitude. This occurs when the wing is at a height equal to 10 percent of its span above the ground. At a height equal to 50 percent of its span above the ground (about 18-feet of altitude), there is only about an 8-percent reduction in induced drag due to ground effect.

## Extent of Ground Effect

A mixture of short runways, rough ground, grass and snow, high altitude, high air temperature, in any of many combinations, is a danger signal.

When you find yourself in a marginal takeoff situation, it is best to know the plane's takeoff speed for the prevailing conditions, the distance required to accelerate to the speed, and then allow a generous margin of safety by picking up as much speed as possible in ground effect before trying to climb. See: 50/70 RULE FOR TAKEOFF.

Twenty to 30 feet up, ground effect vanishes. The wing encounters the full effect of induced drag.

Immediately after liftoff, maintain the wing as close to the ground as is practical while accelerating after a soft-field or short-

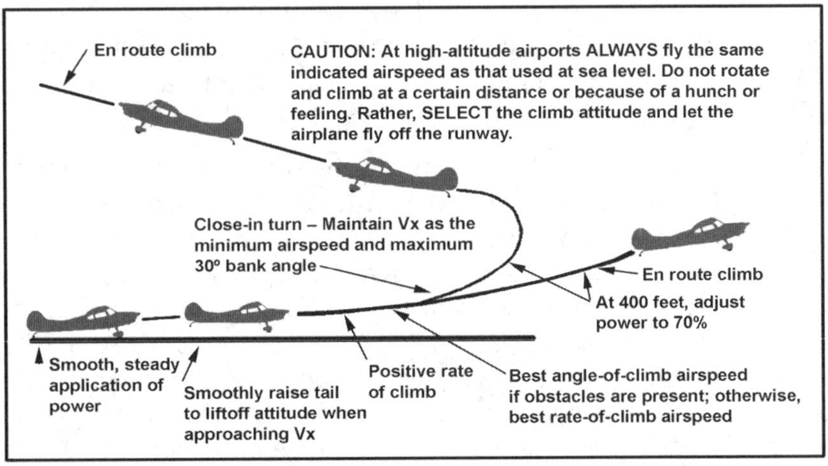


*Figure 4.* **Normal takeoff.**

field takeoff. During the acceleration to $V_X$ or $V_Y$, fly no higher than one-half the wing span above the ground.

## NORMAL TAKEOFF

During a normal takeoff, the airplane becomes airborne smoothly and with positive control. This is done by "rotating" the airplane about its lateral axis to establish a flying *attitude*. It is not accomplished by pulling back on the control wheel until the airplane starts flying. It is much safer to place the airplane in the takeoff attitude and allow it to fly off.

Depending on excess power available, the airplane should be rotated to a nose-up attitude (takeoff attitude) of between three and seven and one-half degrees. This technique has its advantages. If there is frost (polished smooth, of course) on the wings, or the airplane is operated at a high density altitude, the airplane will not fly until it can fly. The technique also compensates for ground effect. By rotating to and maintaining the best rate of climb airspeed attitude (the position of the nose in relation to the natural horizon), the airplane will continue to fly once it is clear of the runway and ground effect.

### Normal Takeoff Procedure

- ❑ *Line up along the centerline of the runway.*

- ❑ *Look down the runway, choosing a prominent landmark at or beyond the opposite end of the runway.*
- ❑ *Make a smooth, steady application of power and maintain directional control. Use full power for takeoffs (use full takeoff power with turbocharged or supercharged engines). Many small airplanes incorporate a "power enrichment valve" in the carburetor for additional engine cooling during the climb. If full throttle is not used, the power enrichener will not activate. Do not over-boost (operate beyond the redline manifold pressure) the supercharged or turbocharged engine. As the elevator control becomes effective, smoothly and slowly apply forward pressure to lift the tail wheel. Rapidly raising the tail will cause a gyroscopic-precession swerve to the left. Raising the tail makes thrust more effective and aerodynamic drag will be at a minimum.*
- ❑ *Raise the nose to lift-off attitude when approaching the best angle-of-climb speed. Rotating too soon or rotating beyond the proper attitude will increase induced drag and delay the takeoff and increase the takeoff distance.*
- ❑ *The airplane will fly off by itself. Maintain attitude while accelerating to $V_X$ (best angle-of-climb speed). Climb at $V_X$ until all obstructions are passed. Then accelerate to $V_Y$ (best rate-of-climb speed).*
- ❑ *Maintain full power until at least 400-feet agl. Full power with the best rate-of-climb speed will result in the greatest altitude gain in a minimum of time. This speed makes an engine failure more managable.*
- ❑ *During gusty conditions, use forward stick to keep the airplane on the ground while accelerating to $V_R$ (velocity of rotation) plus one-half*

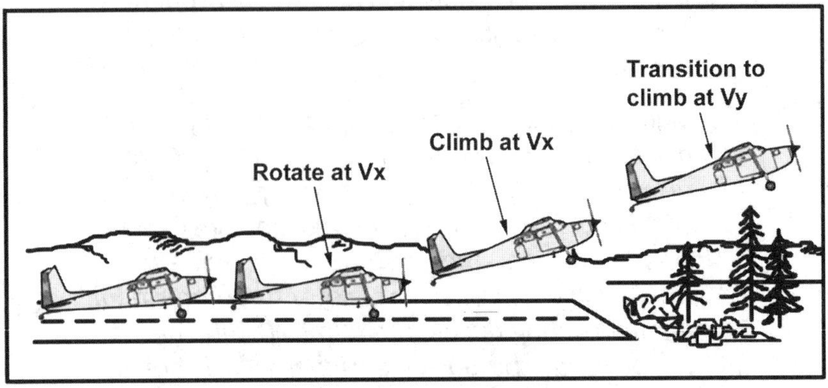


*Figure 6.* **For a short-field takeoff use the "50/70" rule. Mark the halfway point of the runway. If 70% of the speed necessary to rotate is obtained at or before the mid-point of the runway, continue the takeoff.**

*the gust value. Make a positive liftoff and transition to the best rate-of-climb airspeed (best angle-of-climb airspeed if obstacles are present). This will prevent lifting off and touching down again with possible side loads.*

- ❑ *Maintain directional control, re-trim, and climb at $V_Y$ to a safe altitude.*

## YAW LEFT DURING TAKEOFF

Remember the taildragger will have a tendency to yaw to the left during the takeoff run because of the following:

*(1) TORQUE - Every action has an equal and opposite reaction. Have you noticed people in a restaurant trying to get ketchup out of a bottle? They hold the bottle inverted and pound on the base (top) ... and pound on the base. The equal and opposite reaction is not for the ketchup to be squirted out, but for it to retract back into the bottle. (Proper technique is to hold the bottle inverted and use the palm of the other hand to strike the side of the bottle in an upward direction).*

General aviation airplanes, having a clockwise rotation of the propeller when viewed from the cockpit, have a torque reaction to that rotation in a counterclockwise or left direction.

Additionally this causes the left main wheel to be pressed on the ground harder than the right main wheel. The extra drag from friction causes the airplane to yaw to the left.

> *(2) P-FACTOR - Propeller factor (asymmetric thrust of the propeller) results when the right side of the propeller's plane of rotation develops more thrust than the left side causing the airplane to yaw left.*

The conventional gear airplane's longitudinal axis is pointed up into the air during the takeoff run. The prop's angle of attack is still the angle between the chord line of the propeller blades and a line perpendicular to the ground; however, the propeller blades are tilted away from the perpendicular line. This tilt causes a large angle of attack on the right plane of rotation and a small angle of attack on the left plane of rotation.

> *(3) SPIRAL SLIPSTREAM - (Corkscrewing effect of the propeller slipstream) - The slipstream from the propeller strikes the vertical stabilizer and rudder on the left side of the airplane, yawing the airplane to the left.*

> *(4) GYROSCOPIC PRECESSION OF THE PROPELLER - (Asymmetric thrust) - Any spinning object with mass exhibits the properties of a gyroscope. One property is precession, that is, it reacts to an outside force, but the reaction is 90 degrees in the plane of rotation. When the tail is raised with the elevator, the effect is the same as a force in front of the airplane pushing down on the bottom of the propeller's plane of rotation to force the tail up. With clockwise rotation (viewed from the cockpit), 90 degrees in the plane of rotation is where the force acts, yawing the airplane to the left.*

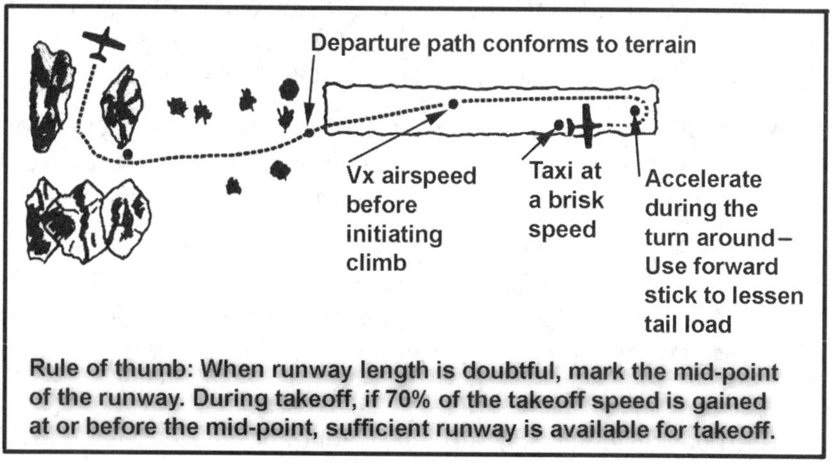


*Figure 5.* Short-field takeoff.

## SHORT-FIELD TAKEOFF

The short-field takeoff is used whenever the field length is limited or there are obstacles near the windward end of the runway or along the departure flight path.

✔ *Flaps should be set as recommended by the manufacturer. The POH or owner's manual will provide information on the use of flaps. If there is no recommendation, make full aileron control deflection, and match the flaps to this deflection.*

Many pilots prefer to lower the flaps just before liftoff when using manual flaps, thereby reducing drag to a minimum and preventing objects from being flung by the propeller blast into the flaps. If the pilot finds it difficult to reach the Johnson bar (floor-mounted flap handle), apply one notch of flaps before beginning the takeoff. It will be much easier to reach the handle for the application of additional flaps. With electric flaps, setting the flaps before the takeoff roll is preferable.

● *Keep in mind some manufacturers advocate the use of flaps for the short-field takeoff only if there are no obstacles present after takeoff. The manufacturer may have determined the use of flaps will shorten the takeoff run, but the rate of climb is reduced with flaps extend-*

*ed to the extent that any benefit of using flaps to shorten the takeoff ground roll is lost in the reduced rate of climb after liftoff.*

- *Make a prompt, smooth application of full power. Do not lock the brakes and add full power (unless necessary to prevent over-boosting on a turbocharged or supercharged engine). The takeoff distance will not be shortened because the propeller must move forward before it becomes efficient. Also, using full power with the brakes locked can pick up small rocks and debris that may damage the propeller or be accelerated rearward to damage the horizontal stabilizer.*
- *As soon as the elevator control becomes effective, use forward control pressure to assume an attitude similar to normal cruise attitude. This will minimize the aerodynamic drag.*
- *Accelerate to the best angle-of-climb speed and rotate to an attitude that will maintain this speed until obstacles are cleared, then transition to the best rate-of-climb airspeed.*

For takeoff from a short-short field, you can gain extra takeoff speed by making the turn onto the runway at a fast taxi speed and by opening the throttle as the plane swings around to line up with the runway. Practice this maneuver with an instructor first to learn how fast this turn may be made without overloading the landing gear. As a rule, the ball of the turn and bank instrument should not exceed one and one-half ball widths during this turn.

All advantage of this turn is lost if the airplane is limping along on one main gear while dragging a wing. The turn should be made into the wind. If required, forward pressure on the control wheel will lighten the load on the tail wheel once the turn begins.

Remember, taxiing with a tail wind will cause a tendency to tip over during the turn due to centrifugal force and the weight of the fuel moving to the outside of the turn.

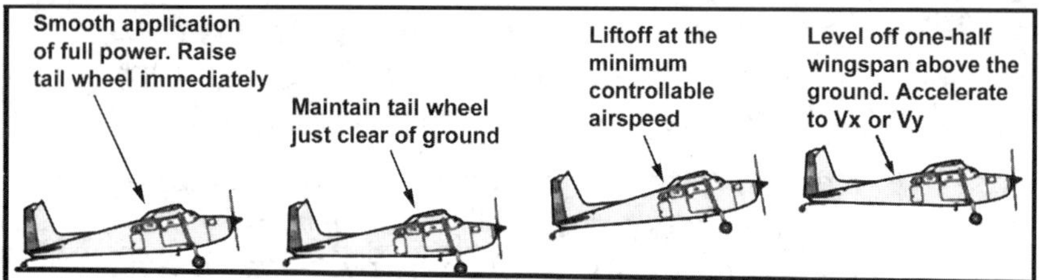


**Soft-field takeoff (soft or rough field of snow, mud, high grass or loose rocks) – The takeoff run is made at a high angle of attack. Raise the tail wheel to barely clear the ground as soon as possible. Liftoff at the minimum controllable airspeed. Reduce the angle of attack and accelerate to Vx or Vy.**

*Figure 7.* Soft-field takeoff.

## SOFT-FIELD TAKEOFF

The soft-field or rough-field takeoff is used when the surface is soft, rough, rock, grass, mud, sand, gravel, snow or slush covered.

The normal takeoff procedure should not be used on a soft field because the surface drag may be sufficient to prohibit takeoff in the distance available.

The object is to transfer the weight of the aircraft from the wheels to the wings as soon as possible. This takeoff procedure assumes that the surface drag of the tires is greater than the induced drag resulting from a high angle of attack.

The tail wheel is lifted off the surface as soon as the elevator becomes effective.

Generally, the soft-field takeoff is made with flaps. If the manufacturer advises otherwise, such as when an obstacle is present, perform the takeoff in the manner prescribed in the flight manual.

- *Smoothly apply full power. Use enough forward pressure to lift the tail wheel as soon as possible, but once it is up, keep the tail as close to the ground as possible. This transfers the weight of the aircraft to the wings rather than the wheels. The induced drag associated with lift is less than the drag of the soft field on the wheels.*

- *The nose-high attitude with the tail wheel just clear of the ground is maintained as the speed increases.*
- *The aircraft lifts off the ground at the slowest possible speed. This airspeed is slower than a safe climb speed and if maintained, the airplane will stall or settle back to the ground when it climbs out of ground effect.*
- *Level off at a height of one-half wing span above the ground and accelerate to $V_X$ or $V_Y$.*

There is no need to climb until the correct speed is attained. This speed is reached faster when the nose is lowered immediately after takeoff so the aircraft can accelerate while just clear of the ground.

When the best angle-of-climb speed is reached, initiate a climb until clear of obstacles, then transition to the best rate-of-climb speed. The sink most pilots expect on retracting the flaps need not occur if there is adequate speed to allow increasing the pitch attitude during the flap retraction.

The soft-field procedure is done to save wear-and-tear on the airplane and to remove the resistance of drag from the main landing gear when operating from a high-drag field. Practice this technique before the need to use it arises.

Some aircraft, those lacking horsepower, require caution. They may be flown into a field where they cannot be flown out. Using the soft-field takeoff technique on a short, rough field may be impossible. In this case, use the modified short-field/soft-field technique.

## MODIFIED SHORT-FIELD/SOFT-FIELD TAKEOFF

To reduce the adrenaline level, but not all the excitement during operations at some short and rough fields, judgment might dictate the use of a modified soft-field and short-field takeoff technique when flying an airplane with low performance.

In an airplane lacking horsepower, the use of a soft-field technique may not allow adequate forward velocity to be obtained fast enough to fly the airplane from the space available. This is because the drag induced with a high angle of attack retards the acceleration of the airplane. Abusing the airplane on a rough field by using a short-field takeoff technique is generally better than using the soft-field technique.

Choose the best available path along the field and unless the field is really soft, rather than rough, the aircraft will accelerate faster with surface friction drag on the wheels instead of induced drag on the wings.

Rather than moving the control stick forward for zero angle of attack, maintain as much lift as possible based on the acceleration of the airplane. The 50/70 rule for takeoff that follows, will allow you to experiment safely to find the optimum angle of attack.

Once the airplane is "hauled" from the ground in a semi-stalled condition, push the nose down to accelerate in ground effect before initiating the climb.

## CROSSWIND TAKEOFF

Crosswind operations are covered in a different part of this book. See Chapter 10 – Crosswind Operations.

## 50/70 RULE FOR TAKEOFF

The takeoff distance for short airstrips can cause confusion and doubt for the pilot. The takeoff distance information provided by the aircraft manufacturer probably doesn't cover vagaries for a runway surface that is grass, sand, gravel or soft turf, at a high altitude, with a tailwind, or if it has a noticeable gradient.

A simple rule of thumb can be used to decide that the runway length is sufficient for takeoff.

### Rule of Thumb - Determine Runway Length Sufficient for Takeoff

*Multiply 10 times the square root of the percentage of liftoff distance required. This equals the percentage of liftoff speed needed in that distance.*

Wow! That sounds complicated; and I promised no unreasonable math computations. To keep that promise we have simplified the rule in the following paragraphs.

The rule of thumb states:

> *Multiply 10 times the square root of the percentage of liftoff distance required.*

Since we are using the half-way point of the runway, this is 50 percent. The square root of 50 is 7.07, which when multiplied by 10, equals 70.7 percent.

The 70.7 percent equals the percentage of liftoff speed that needs to be attained at the halfway point of the runway to guarantee that the airplane will takeoff in the space remaining.

If 70.7 percent of the takeoff speed is not attained, abort the takeoff. In this case you will have to wait until conditions change, that is, the temperature cools, the wind changes or off-load some of the weight.

Remember, this rule does not guarantee that the rate of climb after takeoff will be sufficient to clear obstacles, but it does guarantee the takeoff can be made in the length of the runway.

The rule automatically compensates for density altitude and sloped runways and other variables.

Once airborne climb to the desired altitude and prepare for the cruise phase of flight.

On runways—especially those where either end is not visible from either end—pace off the length to determine the halfway point and mark it physically or mentally.

Using the halfway point of the runway simplifies the rule and provides a safety margin. Most airplanes stop better than they accelerate. For example, it is possible to land an airplane at an airstrip where the length is not adequate for takeoff. The landing distance is a lesser value than the required takeoff distance.

This rule of thumb will show if you can takeoff. It is not necessary to make lengthy or complicated calculations and then worry about whether or not you have considered all variables. If, during takeoff using this rule of thumb, the required speed is not attained at the halfway point of the runway, there is still adequate stopping distance available.

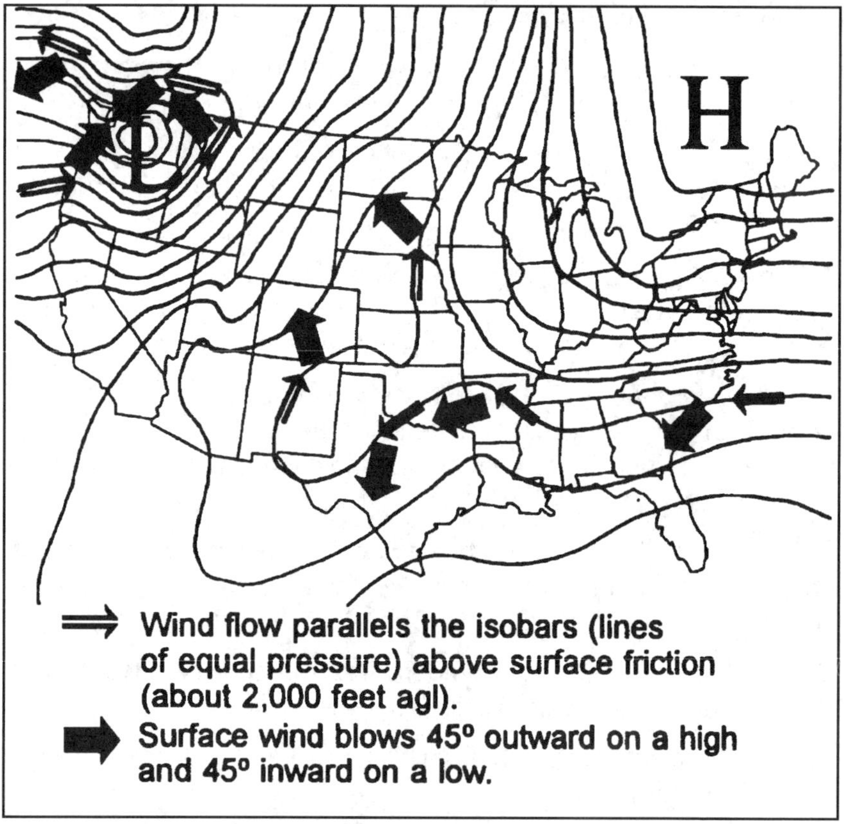


*Figure 8.* Predict the general weather trend by facing the surface wind. Extend your left arm 45 degrees to the left and the right arm 135 degrees to the right. This is 180 degrees total, a straight line.

The left hand points to a high-pressure area. The right hand points to a low-pressure area.

The intensity of any incoming weather (low pressure area) cannot be determined by this method, but an idea of the general weather (better or worse) may be predicted.

# CHAPTER 6

# Cruise

## ENGINE OPERATIONS

This section on engine operations is concerned with the fuel-air mixture ratio and the procedure for leaning the mixture.

### Mixture Control

The mixture control enables the pilot to adjust the fuel-air mixture for more efficient engine operation, fuel economy and flight safety.

Improper operation of the mixture control causes engine problems with the possibility of engine failure. Leaned more than the manufacturer's recommendation, the engine runs rough with sudden "cutting out," backfiring, detonation, overheating and loss of power. Enrich the mixture excessively and you can expect rough engine operation, loss of power and poor economy. Often it is impossible to hear detonation in a noisy cockpit or to feel it because of the normal vibrations of ground and flight operations.

Do not lean turbocharged or supercharged engines for takeoff. These engines provide sea-level power up to their critical altitude and they need sea level fuel flow.

A fuel-air mixture ratio of 1 to 15 (.067) is the chemically correct (stoichiometric) mixture for the fuel burn to be considered optimum. This mixture setting provides the maximum fuel utilization, that is, the maximum energy output per unit of fuel.

Lean the mixture beginning from the full-rich mixture position. The exhaust temperature increases because the amount of excess fuel is decreased. Using an exhaust gas temperature gauge (EGT), the mixture is leaned (pulled or screwed out) until there is no excess fuel, nor air. This is the "peak EGT" temperature or highest temperature that can be obtained.

Even though engine operation at peak EGT is considered the chemically correct mixture for combustion, it produces the greatest heat and causes the engine to run at its hottest.

Look at figure 1 and examine the cylinder head temperature at the .067 mixture setting. Next, look at the mixture setting range of .076-.082. The ratio of 1 pound of fuel to 12½ pounds of air is known as the best power mixture setting. The cylinder head temperature is less than the best economy mixture setting (.067) and the engine produces its greatest power.

**Always remember you fly an airplane with your head, not your hands.** – *Anonymous*

## Fuel-air Mixture

Gasoline engines require approximately 15 pounds of air to burn one pound of fuel completely. The fuel-air ratio is the proportion between the weight of the fuel and the weight of the air that goes into the cylinders.

The mixture of fuel and air depends on the carburetor, unless the engine is fuel injected. When fuel passes through the carburetor, air is mixed before it enters the cylinders. Gasoline cannot ignite when in the liquid state. It must be atomized or vaporized (broken into minute droplets) by the carburetor.

When the airplane climbs to an altitude above sea level, the volume of air entering the carburetor remains the same, but the weight of the air decreases as the air density decreases. This changes the fuel-air weight ratio requiring adjustment with the mixture control.

The amount of fuel (by weight) entering the carburetor depends on the volume of air (not the weight) passing through the throat of the carburetor. Without adjustment, as the altitude increases and the weight of the air is less (the volume remains the same), the mixture becomes richer. It becomes important to change the fuel-air mixture with the mixture control. The mixture control compensates for the decreased air density by metering the amount of fuel that passes through the main jet in the carburetor.

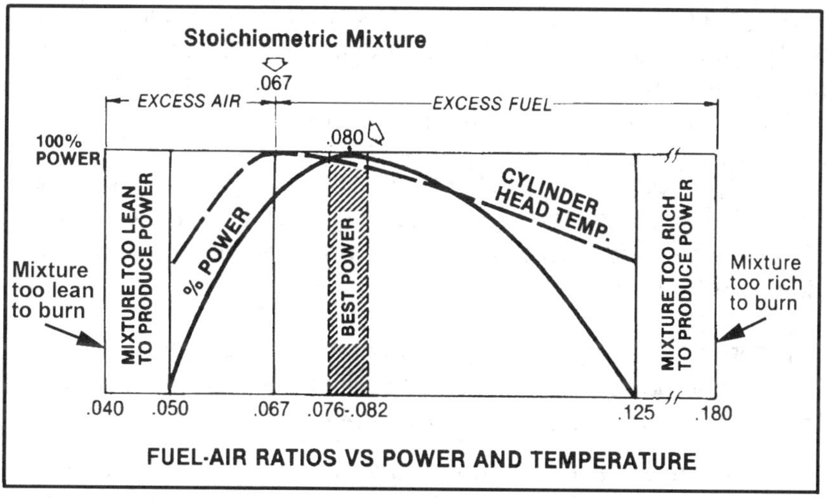


*Figure 1.* **The stoichiometric (chemically correct mixture to burn all the fuel the engine is capable of burning) mixture results in the greatest cylinder head temperature and a loss of power. The best power mixture yields the greatest power with slightly cooler cylinder head temperature.**

## Temperature and Fuel-air Ratio

The fuel-air ratio of 1:15 (one to 15 or .067) is the stoichiometric mixture (chemically correct) where the fuel burn is normally considered optimum because it is the mixture setting that gives the maximum efficiency of fuel use. This is the maximum energy output per unit of fuel.

An airplane equipped with an EGT (exhaust gas temperature) gauge takes the guess work from leaning the mixture. As the mixture is leaned from the full rich position, the exhaust temperature increases because excess fuel is being decreased. When the mixture control is pulled out (leaned) until there is no excess fuel or air, the highest temperature is recorded. This is the "peak EGT." If leaned beyond this point, the temperature will begin to decrease.

Though engine operation at peak EGT is considered the chemically correct mixture for combustion, it causes the engine to run at its hottest. ALCOR, a manufacturer of EGTs and combustion analyzers, recommends 65-percent power as the maximum for operation at peak EGT. The author has always felt that although some engine manufacturers allow operation at peak EGT at power settings more than 65 percent, ALCOR's advice is sound. Buying a little extra gas is cheaper than buying engine parts. For cruise power settings above 65 percent power, the fuel-air ratio of 1:12½ (.080) is used and is known as the "best power" mixture.

Teledyne Continental has not approved operating at peak EGT for cruise at power settings above 55-percent power (but 65 percent may be used with some engines provided certain RPM limitations are followed).

Lycoming has approved operating at peak EGT for cruise up to 75-percent power (except geared-supercharged engines, where they set the limit at 65-percent power).

Check the POH (pilot's operating handbook) for the approved operations concerning your particular engine. There have been changes in recent years, that increase safety and performance, that allow operation on the lean side of peak.

### Recommended Leaning Altitude

The old timers will remember when engine manufacturers recommended that the mixture should not be leaned below 5,000 feet. The reason was that below this altitude there is no noticeable increase in engine efficiency. At 5,000 feet and higher, the normally aspirated engine can develop a maximum of 75 percent or less of its rated power. At the higher altitudes the pilot is less likely to get into trouble if he uses an improper leaning technique.

The FAA now states the mixture can be leaned at any altitude providing you adhere to the proper procedure. The altitude you use for leaning the mixture should be based on the density altitude, not the physical altitude. The density altitude is the altitude at which the airplane "thinks" it is operating and it will perform in accordance with this density altitude.

There is a trend to better educate pilots on the procedures and hazards of using the mixture control. Armed with this information, the pilot should lean the mixture whenever the cruise power setting is 75 percent or less, at any altitude.

## Mixture Distribution

The term "perfect mixture distribution" refers to a condition that is highly desirable, but unfortunately, impossible to achieve. Mixture distribution refers to the quantity of fuel-air mixture delivered to each cylinder.

Most engine designs use a carburetor mounted at the rear end of the engine. When air enters the carburetor, fuel is mixed to become the fuel-air mixture. This mixture enters the intake manifold. Those cylinders closest to the carburetor receive the richest mixture. Those cylinders at the front of the engine—farthest from the carburetor—are the leanest. There is no uniformity in the fuel-air mixture delivered to each cylinder. The mixture must be leaned according to the leanest cylinder to prevent engine damage. This results in too much fuel going to some cylinders, but it is unavoidable.

In an attempt to increase performance and economy, manufacturers developed fuel-injection engines for improved control of the fuel-air ratio and more uniform delivery to each cylinder.

The carbureted engine uses the pressure differential of airflow through the carburetor venturi to distribute the fuel-air mixture to the cylinders. The fuel injection system uses a fuel-injection pump as the pressure source to introduce the fuel to the intake port of the cylinder just ahead of the intake valve, or directly into the cylinder combustion chamber.

The typical fuel-injection system has a fuel injector pump to separate vapor from the fuel. The fuel is then pumped to a fuel-air control unit that replaces the carburetor in obtaining the proper fuel-air ratio. This fuel-air mixture then passes through a fuel manifold that contains outlet ports for each cylinder. Lines from these outlet ports direct fuel/air mixture to individual discharge nozzles at the cylinder intake port. Although this system approaches perfect mixture distribution, any restrictions in the manifold valve, outlet ports, lines or discharge nozzles can cause a cylinder to become suddenly lean.

### Detonation

Over leaning the fuel-air mixture is a major factor contributing to fuel detonation in the cylinders. Others are incorrect octane rating of the fuel, excessive cylinder head temperatures, excessive manifold pressure, incoming mixture temperature and an abrupt opening of the throttle.

Detonation is a destructive, spontaneous explosion of the unburned gases in cylinders after normal ignition. It happens when the temperature and pressure of the unburned fuel reach critical values. The result is a sudden and violent explosion at around 1,500 miles-per-hour, rather than the relatively slow, smooth burning of normal combustion at around 34 mph.

Usually detonation cannot be recognized from the cockpit through sound or engine roughness. Protection against its possible occurrence must be strictly followed.

Continued operation when detonation is present can result in dished piston heads, collapsed valve heads, broken rings, or eroded portions of valves, pistons, or cylinder heads. Detonation may end with the sudden and complete failure of the engine.

### Preignition

Preignition is the uncontrolled firing of the fuel-air mixture before normal ignition. It is commonly caused by an incandescent area in the combustion chamber that fires the fuel/air mixture prematurely. A glowing spark plug, incandescent exhaust valve or glowing particles of carbon or lead can cause preignition.

Preignition presents a special hazard on the ground after an engine has been run at cruise rpm for a prolonged time or during ground run-up for testing.

Though the magneto switches and fuel are turned off, handling the propeller before the engine cools can be extremely dangerous. Sufficient unburned fuel may remain in the cylinders to cause the engine to kick over a few revolutions if the propeller is moved enough to create compression on the cylinders. Preignition can be as destructive as detonation or worse.

## MIXTURE DEFINITIONS

### Best Economy Mixture

The best economy mixture results when the exhaust gas temperature gauge (EGT) is adjusted to the "peak EGT." Peak EGT is obtained by leaning the mixture until the EGT needle peaks and then begins moving down on the scale. Enrich the mixture until the peak value is obtained again. At this setting there is a small loss of true airspeed over that obtained from the "best power" mixture setting. The range is increased about 10 to 15 percent over the best power mixture setting. Follow the engine manufacturer's specific recommendations for leaning.

### Recommended Lean Mixture

Adjust the mixture to peak EGT, then enrich 50°F for the recommended lean mixture. The recommended lean mixture setting results in a slight true airspeed loss and slight range increase (about five to seven percent) from the best power mixture setting.

### Best Power Mixture

The best power mixture setting results in the maximum speed. Adjust the EGT to peak, then enrich 100°F.

## PROCEDURE FOR LEANING

The following is the traditional thinking concerning the procedures to be used for leaning the mixture.

### Takeoff Mixture Adjustment

With a normally aspirated engine, use full power for takeoff at all airport elevations. Lean the mixture to avoid engine roughness or power loss at airports above sea-level density altitude.

Some pilots prefer to lean to the maximum RPM while operating at full throttle, before takeoff. If this technique is used, limit the engine operation at full throttle on the ground to a minimum.

Unless the pilot cannot chew gum and pat his belly simultaneously, there is a technique preferred to leaning at full throttle before takeoff. This technique works for the normally aspirated engine equipped with a fixed-pitch or a constant-speed propeller. It works for airplanes with or without an EGT gauge or fuel-flow gauge. Begin the takeoff roll, advance to full throttle and ease the mixture control out until the engine starts to run rough. Enrich until the engine runs smoothly.

Do not lean turbocharged or supercharged engines for takeoff. These engines provide sea-level power up to their critical altitude and require sea-level fuel flow.

## CRUISE POWER MIXTURE ADJUSTMENT

### Tachometer Method—Fixed-pitch Propeller

After establishing the cruise power setting, lean the mixture until there is a slight loss of RPM. Enrich the mixture slowly until the RPM peaks at its highest value, then enrich slightly.

### Engine-rough Method—Constant-speed Propeller

What is the best method to lean the mixture on an airplane equipped with a constant speed propeller, but without an EGT or fuel-flow gauge? For cruise flight, adjust the RPM and throttle as desired. Then lean for engine roughness and enrich until the engine runs smoothly.

### Fuel-flowmeter Method—Constant-speed Propeller

For an airplane equipped with a constant-speed propeller, having a fuel flow gauge, but no EGT gauge, set the mixture according to the POH to the proper fuel flow for the particular density altitude.

### Pressure Carburetor

An airplane equipped with a pressure carburetor has an automatic mixture control. Do not manually lean this engine.

### Recommendation

John Deakin has written a 4-part series on *Where should I run my engine?* The articles can be found at www.avweb.com and they provide in-depth science for leaning the engine.

The next part will deal with turbulence and maneuvering speed. An understanding will prevent 'breaking' the airplane.

## TURBULENCE

One would think that with the advancement of knowledge, technical equipment and experience, the science of meteorology could forecast, with accuracy, the occurrence of turbulence. Although certain guidelines have been established for general forecasting, the exact extent and severity of turbulence cannot be pinpointed. Engineers, however, can predict how much stress is necessary to cause structural failure of the aircraft from turbulence. It behooves pilots to become familiar with turbulence and the recommended actions to follow during such encounters.

In mountainous areas, the amount and location of turbulence and up-and-down drafts are predicted mostly by wind velocity and direction near ten thousand feet. Sharp ridges perpendicular to the wind flow are the most dangerous. Due to funneling and venturi effect, the wind over any particular area may vary from the forecast wind.

Three major classifications of turbulence exist: convective currents, mechanical turbulence and wind shear. To combat them all, slow to the maneuvering speed when their intensity causes moderate or greater turbulence.

### Convection Currents (Thermals)

As a parcel of air near the earth's surface is heated, it is forced upward—like a hot air balloon—through a colder environment. Because of the instability created in the atmosphere by the evacuation of the convective area, in regaining stability there is a down rush of air creating a compensating downdraft. These downdrafts generally occur over a broader area than the updrafts.

Convective currents causing turbulence at low altitudes are a source of annoyance. The updrafts persist to an altitude where the temperature of the updraft air, through expansion and cooling, reaches that of the surrounding air. With sufficient moisture, cumulus clouds appear. Thermal tops are generally around 10,000 to 12,000 feet; however, in the mountains, tops extending beyond 15,000 feet are not unusual.

Flying in convective turbulence can be reduced with an early start in the morning, or flight above the convective clouds should eliminate it completely.

### Mechanical Turbulence (Obstruction to Wind Flow)

The smooth flow of wind being disrupted by rough ground or obstructions creates a random motion known as "eddies." In a light wind, these rotating eddies form pockets of irregular whirls to the windward and leeward side of obstructions. If the wind exceeds about 20 knots, these random fluctuations are carried downstream with the wind flow.

While increasing wind velocity produces less variability in direction, it lends to a greater variability in speed. This change in speed and direction may be accompanied by brief, sudden, irregular periods of low speeds (lulls) and high speeds (gusts). The "gust" is a small-scale, sudden and brief increase in the speed of air movement. Gusts are reported by the weather people only if at least nine knots variation occurs between peaks and lulls.

Gusts can be numerous, irregular, random, sudden and brief turbulent motions. Approaching mountains, they can swirl up, slow and cause unpredictable winds, updrafts and downdrafts.

When approaching a ridge, this and other types of turbulence will provide advanced warning before one crosses over. If this turbulence is encountered when flying toward or into mountainous terrain, it is almost sure to get worse farther on. If one decides to continue, approach the ridges at a 45-degree angle to allow an escape route toward lowering terrain with a minimum of maneuvering.

### Wind Shear

Wind shear is best described as a change in wind direction and/or speed in a very short distance in the atmosphere. When two air currents flow past or over each other in different directions or at different speeds they establish a "shear zone" of irregular motion in vertical or horizontal dimensions.

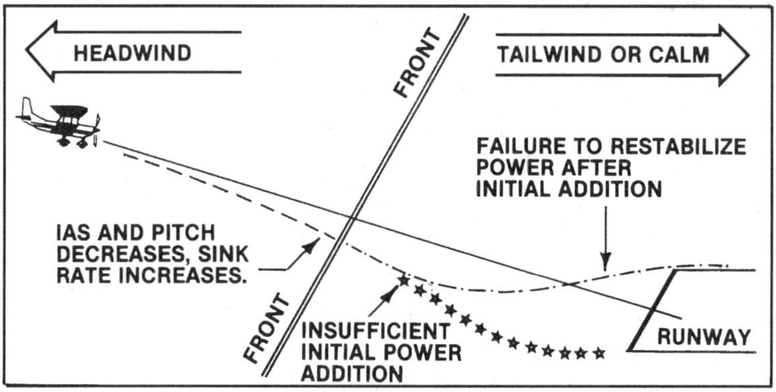


*Figure 2.* Headwind shearing to a tailwind or calm wind condition.

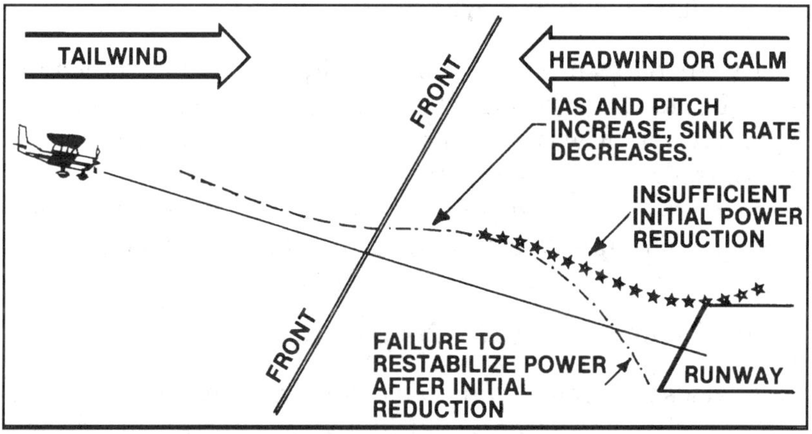


*Figure 3.* Tailwind shearing to a headwind or calm wind condition.

If the winds aloft forecasts report wind 2720 (270° at 20 knots) at 9,000 feet, and 2941 (290° at 41 knots) at 12,000 feet, the velocity changes 21 knots in 3,000 feet. This results in shear of seven knots per thousand feet.

Overall, six knots per thousand feet is sufficient shear to produce turbulence. Obviously, the higher the shear value and the greater the difference in wind flow direction between the two currents, the stronger the turbulence becomes.

The more important shear is that of vertical dimensions that can cause changes in the angle of attack. Conditions associated with shear can occur at any level in the atmosphere, but three are of special interest to pilots.

### Low-level Temperature Inversion

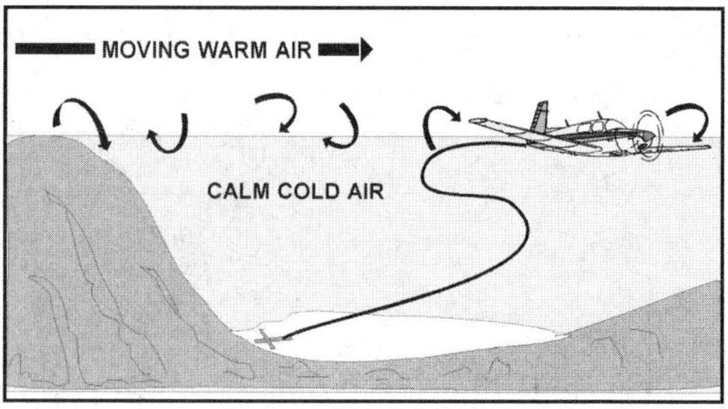


*Figure 4.* **Low-level temperature inversion.**

Clear, cool nights with little or no wind may cause the ground to cool rapidly, cooling a layer of air next to the ground faster than the air several hundred feet above. This results in a temperature increase with altitude, or an inversion of the normal lapse rate. Essentially, a lid is placed over the region. Any wind flow above this layer creates shear.

For years pilots have been told an airplane on the ground may be subject to the wind plus the relative wind due to the motion of the aircraft. It was explained that once the airplane left the ground, the wind would not affect flight other than creating drift and changing the groundspeed. This is not true if the wind changes faster than the aircraft's mass can be accelerated or decelerated.

### Wind Shear in a Frontal Zone

The winds in a frontal zone can be, and usually are significantly different, changing abruptly, in the two air masses that meet to form the frontal zone boundary. How much turbulence depends on the magnitude of the wind shear. While the wind flow direction above and below a front can be accurately determined, the extent (height) above the airport cannot be measured. We can, however, make valid predictions for the approximate height of the wind shear associated with various fronts.

The cold front slope extends rearward from the direction of movement as a wedge of air, wind shear occurs just after the front passes an airport and for a short period afterwards. If the front is moving at 30 knots or more, the frontal slope will usually be 5,000 feet above the airport about three hours after the frontal passage.

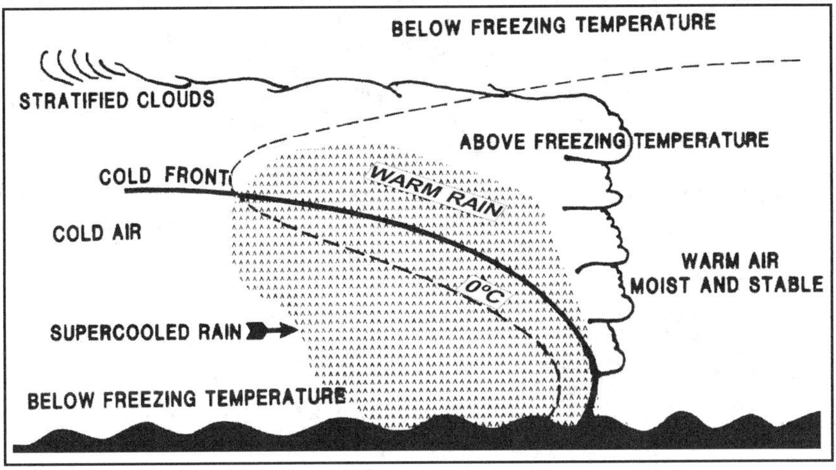


*Figure 5.* **Cold front.**

With the warm front, the slope extends forward in the direction of movement (overriding air). The slope is shallower than a cold front and it moves slower. The most critical period is before the front passes the airport. Warm-front shear may exist below 5,000 feet for approximately six hours. The problem ceases to exist after the warm front passes the airport. The amount of shear in warm fronts is much greater than that found in cold fronts.

Turbulence may or may not exist in wind shear conditions. Broken clouds usually suggest a broad frontal zone where only light turbulence is present. If the surface wind under the front is strong and gusty, there will be some turbulence associated with wind shear. If the horizontal shear is more than 15 knots, or the vertical shear exceeds six knots per thousand feet, one can expect turbulence. Cloud formations with sharp edges show moderate or greater turbulence.

Trough lines extend vertically and horizontally over a large area. With an eastward movement of around 20 knots, expect light turbulence; around 30 knots, expect moderate turbulence.

## Mountain Wave Wind Shear

The mountain wave often creates low-level wind shear at airports downwind of the wave. ACSL (altocumulus standing lenticular) clouds either observed or reported, are clues that shear may exist. Airports downwind from mountain ranges that have a gradually sloped lee side are more suspect since the wind can be accel-

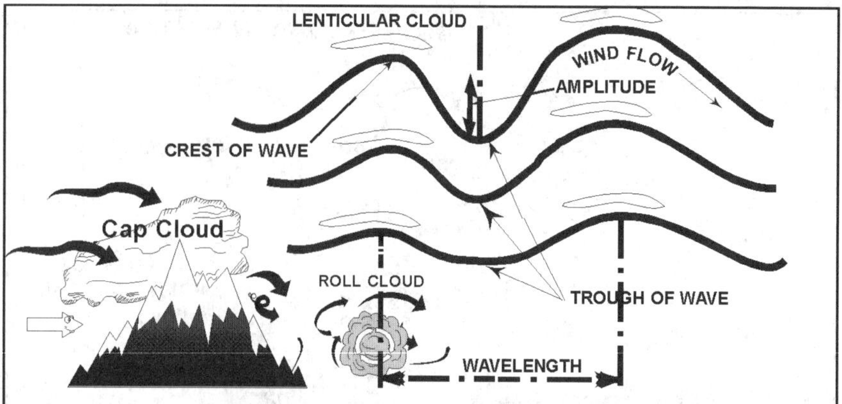


CROSS-SECTION OF THE MOUNTAIN WAVE

*Figure 6.* Mountain wave.

erated downslope, creating greater turbulence of the destructive variety.

## Sea Breeze Fronts

Temperature differences between land and large bodies of water can create local airflows. Convection causes the air to rise over the warmer surface. Air from the cooler surface rushes in and fills the area evacuated and sets up the airflow pattern. The changes in wind velocity and direction can occur in relatively short distances. Airports situated near large lakes, bays or oceans are subjected to this shear.

## Clear Air Turbulence (CAT)

CAT may be associated with a jet stream or strong mountain wave circulation. The name implies turbulence at high levels without clouds. CAT is most frequently generated on the polar side of the jet stream or north and northeast of a rapidly deepening surface low. Mountain wave CAT may extend from the mountain crests to as high as 5,000 feet above the tropopause and can range 100 miles or more downstream from the mountains. This occurs from southern California, across the Rockies and into the Central Plains.

It is best to avoid areas where horizontal shear exceeds 40 knots within 150 miles or where vertical shear exceeds six knots per 1,000 feet.

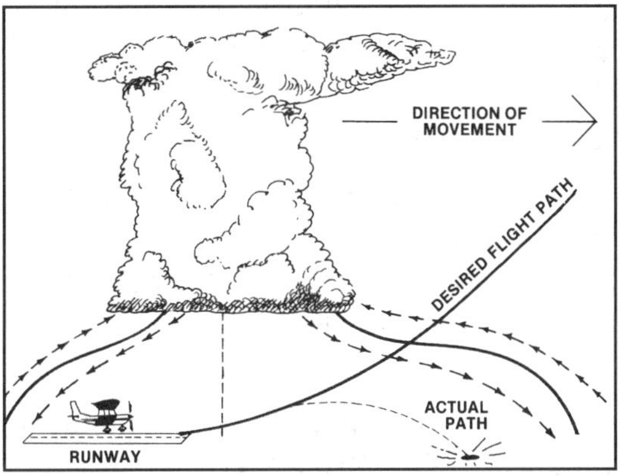


*Figure 7.* Thunderstorm shear.

## Thunderstorms

Thunderstorms may be generated in the presence of moist, unstable air when there is some type lifting action. The altitude where air becomes unstable is called the "level of free convection."

Anytime air moves upward, it expands because of decreasing atmospheric pressure. Conversely, when air moves down, it is compressed by increasing atmospheric pressure. When the pressure and volume change, the temperature also changes. When air expands, it cools; when compressed, it warms. These changes are adiabatic, meaning no heat is removed from or added to the surrounding air.

Unstable air displaced upward, though it cools during ascension, may remain warmer (lighter) than the surrounding air. The lapse rate of the atmosphere here decreases at a faster rate than the air being displaced upward. The rising air continues upward by itself when the lifting action is removed.

The lifting action that starts the process can be promoted by a front, mountainous terrain, convection, convergence or any combination.

When the three conditions of moist, unstable air and lifting action are combined, the thunderstorm is born. The winds around a thunderstorm are complex. Wind shear can be found on all sides of the cell and in the downdraft directly under the cell. The wind shift line (first gust or gust front) can precede the actual storm by 15 nautical miles or more. If a thunderstorm is near an airport of intended takeoff or landing, low-level wind shear hazards may exist.

### Detecting Wind Shear

Neither pilots nor their airplanes can fly through all intensities of wind shear, low-level or high-level. It is necessary to learn to detect, predict and avoid severe wind shear. Severe wind shear does not strike without warning. It can be detected by the following methods.

Analyze the weather during preflight. If thunderstorms are observed or forecast near the airport, be alert for the possibility of wind shear in the departure or arrival areas. Check the surface weather charts for frontal activity. Determine the surface temperature difference immediately across the front and the speed the front is moving. A 10°F or greater temperature differential, and/or movement of the front at 30 knots or more provides warning of significant low-level wind shear.

PIREPs should be specific, giving the loss or gain or airspeed due to shear and the altitude or altitudes where it was encountered. Areas with reported shear that causes airspeed losses more than 15 knots should be avoided. Thunderstorm movement can cause one aircraft to encounter an airspeed increase, that is harmless, whereas the next aircraft can experience a severe airspeed loss.

Assume severe wind shear is present when there is a combination of any two or more of these conditions:

- ✔ *Extreme variation in wind velocity and direction in a relatively short time.*
- ✔ *Evidence of a gust front such as blowing dust or debris on the airport surface.*
- ✔ *Surface temperature more than 80°F.*
- ✔ *Dew point spread of 40°F or more.*
- ✔ *Virga - precipitation aloft falling from the bases of high-altitude cumulus clouds that evaporates before reaching the ground.*

If frontal activity exists, note the surface wind direction to figure out the location of the front with respect to the airport. All fronts produce a 90- to a 180-degree clockwise wind shift with their passage. If the airplane will traverse the front, compare the surface wind direction and speed with the wind direction and speed above the front to detect the potential wind shear during approach or landing.

With DME, LORAN or GPS, the pilot can compare the true airspeed and groundspeed. Any rapid change represents a wind shear. Compute a minimum groundspeed for the approach by subtracting the surface head wind from the true airspeed.

With DME, the ILS or VASI can give an indication of shear. Even for the pilot that is not instrument rated, there is nothing wrong or illegal about following the glide slope of an ILS in VFR weather. If VASI is available it will provide the same information.

The rate of descent is dependent on ground speed. At the normal approach speed, following a glide slope (or glide path), if a high rate of descent is required to maintain the glide slope, this shows a strong tail wind. A low descent rate denotes a strong head wind.

In a potential shear situation, power will vary also. Less power than normal affirms a tail wind; more power indicates a head wind.

Pitch attitude is also an indicator. A lower pitch attitude than normal shows a tail wind. A higher pitch attitude means a head wind.

By observing these parameters, one gets a feel for the wind velocity. The wind correction angle shows the wind direction. These can be compared with the reported surface winds for an index to gauge shear before its encounter.

## AIRPLANE PERFORMANCE IN WIND SHEAR

### Power Compensation

When an aircraft travels rapidly from one air current to another, it undergoes abrupt changes in acceleration.

Assume an aircraft, climbing or descending, encounters an instantaneous wind shear where the airplane was experiencing a head wind and the shear is to a calm or tail wind condition. The shearing may result from a temperature inversion, crossing a frontal zone or encountering mountain wave shear. At this instant several things will happen: the airspeed will drop an amount equal to the difference between the head wind and calm/tail wind velocity, the nose will pitch down and the aircraft will begin to lose altitude. The airplane is in a "power deficient" state. If the nose is pulled up to a point even higher than before the shear to maintain altitude, the airspeed loss is further aggravated. On a landing approach, thrust must be increased to create acceleration. If the airplane is too slow and too low, a short landing results.

With a tail wind shearing to a head wind or calm wind, the reverse occurs.

### Energy Trade

There are only two ways an aircraft can correct for a wind shear. There can be an energy trade or a thrust change. Most small taildraggers do not have an excess of power available, and that brings us to the energy of motion. Rather than go through the proof of mathematical equations, we can say that kinetic energy is directly convertible to energy of vertical displacement (potential energy). More simply put, airspeed can be traded for altitude or vice versa.

The pilot caught in low-level wind shear who finds the airspeed is slower than normal and approaching a stall, although full power has been applied, has but one option. Lower the nose and regain speed by trading away altitude. This is trading potential energy for kinetic energy. It is necessary, but the penalty for doing this is severe. A large sink rate is built up and a great deal of altitude is lost for a relatively small increase in airspeed. At a low altitude, maintaining a stabilized lower airspeed may be preferable and rely on the airplane's climb performance rather than pushing the nose over and risking ground contact.

## FLIGHT LOAD FACTORS

An airplane's strength is measured by the total load the wings can carry without distortion or permanent damage. The load imposed upon the wings depends mainly upon the type of flight. The wings must support not only the weight of the airplane, but also the additional loads of centrifugal force imposed during maneuvering or when subjected to turbulence.

In straight-and-level flight, the wings support a weight equal to the weight of the airplane and its contents. If the airplane is moving at a constant airspeed in a straight line, the load on the wings remains constant (assuming no turbulence). When the airplane assumes a curved flight path or pulls out from a dive, the actual load on the wings will be greater because of the centrifugal force imposed during the period of curved flight. The additional load results in stresses in the wing structure.

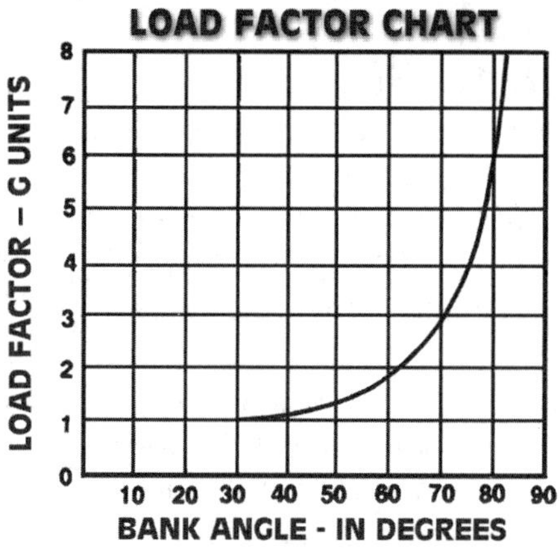


*Figure 1.* Load factor chart.

The load factor is the actual load on the wings divided by the weight of the airplane and its contents. To effect a turn, increased lift pulls the airplane from its straight course while overcoming gravity. As the bank becomes steeper, the total lift must increase and centrifugal force builds up.

A positive flight load factor is one where the aerodynamic force acts upward with respect to the airplane. With a negative flight load factor, the force acts downward.

One additional cause of large load factors is severe vertical gusts. These gusts cause a sudden increase in the angle of attack resulting in large wing loads resisted by the inertia of the airplane.

For certification in the normal, utility or acrobatic category, the aircraft's structure must be able to support the limit loads specified without detrimental, permanent deformation. At any load up to the limit load, a deformation may occur, such as the wings moving up and down, as long as it does not interfere with safe operation of the airplane and the deformation is not permanent.

The G load factor is not caused by the aircraft's gross weight, or its airspeed. When the aircraft assumes a circular path, centrifugal force increases. Any airplane, despite its size, gross weight or airspeed, that assumes a coordinated level flight, 60-degree banked turn, will experience a 2-G load factor.

## Ultimate Loads

Additionally, the structure must be able to support ultimate loads without failure for at least three seconds. FAR Parts 23 and 25 require the ultimate loads to be 50 percent greater than the limit load factor. This safety factor is to compensate for construction irregularities and differences in materials. It does not allow the pilot to intentionally fly above the limit load factor.

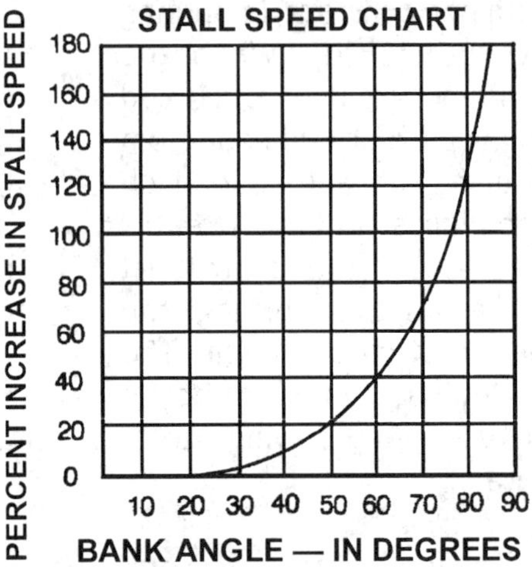


*Figure 2.* **Stall speed chart.**

## MANEUVERING SPEED

The newer owner's manuals and POHs (Pilot's Operating Handbooks) show the maneuvering speed changes with a change in gross weight. This might be opposite of what one would expect. As the weight decreases, the maneuvering speed decreases.

Maneuvering speed, sometimes called "rough air speed," is the maximum speed where full abrupt control deflection can be made without exceeding the design load factor. Maneuvering speed allows the airplane to stall, relieving pressure, before it experiences structural failure.

We know the maneuvering speed decreases in value with a decrease in gross weight. Pilots know a heavily loaded airplane rides smoother in turbulent air. Because of this some pilots perceive this as an indication the airplane should be loaded to its maximum allowable gross weight when turbulence is expected. Should they?

It is true that the inertia of a heavy airplane will resist vertical motion greater than a lighter airplane. Because of this less Gs are produced. However, the stress imposed on the airframe is a more important consideration.

During flight the pilot recognizes load factors; the aircraft structure recognizes only the total load imposed on its structure.

Consider one airplane encountering a fixed gust condition at two different gross weights. We will use a gust of +30 fps (feet per second), because a regulatory requirement of all airplanes is that they must be capable of withstanding a +30 fps gust at maximum level-flight speed at normal rated power.

The airplane is first loaded to 3,800 pounds. It encounters the +30 fps gust that results in an arbitrary additional 2-G load factor. The airplane, off-loaded to 1,900 pounds, encounters the same gust. The acceleration due to the gust is an additional 4-Gs load factor (with half the inertia, the gust acceleration is double).

The heavy airplane has one G in normal flight plus two Gs from the gust. Multiply the three Gs by the gross weight, 3,800 pounds, for a total load supported by the wings of 11,400 pounds.

The light airplane has one G in normal flight plus four Gs from the gust. Multiplying the weight and force results in a total load of 9,500 pounds.

The lightly loaded airplane is subjected to 1,900 pounds less load when encountering the same gust. The heavy plane, realizing less load factor, incurs more strain.

If one flies at the maneuvering speed when operating in areas of moderate or greater turbulence, there is no worry about the aircraft's structure.

### Rule of Thumb: Determine the maneuvering speed

*Many older taildraggers do not have a POH or even an owner's manual. If the maneuvering speed cannot be determined for an older airplane, use the value of 1.7 $V_{SO}$. Multiply the normal stall speed ($V_{SO}$) by 1.7. For some airplanes the published maneuvering speed may be as much as twice the stall speed, but one will not break anything by using the value of 1.7.*

## Increased Stall Speed

The stall speed increases while turning. Because the load factor increases with the resultant centrifugal force of the turn, a greater amount of lift (a larger angle of attack) is required. The critical angle of attack is attained at a higher speed during a turn.

The stall speed increases as the square of the load factor. In a 60-degree banked turn the load factor is two Gs. The square root of two is 1.41, or a 41 percent increase in stall speed.

## EMERGENCY LANDING

Statistically you are pretty safe when it comes to an unexpected engine failure that leads to a forced landing. But, when an engine quits, statistics aren't too comforting.

### Energy Absorption

It is important to use whatever is available to cushion the forced landing. This may be the aircraft structure or environmental materials (vegetation, trees, man-made structures, crops, and the like). A word of warning. If you need to land in trees, pretend the tree tops are an international airport and make a normal landing to the tops. Pilots who slow down and stall above the tree tops often experience compression force. The human body can withstand something like 4-5 g's of compression force and 40-50 g's of lateral force.

### Occupant Restraint

Seat belts and shoulder harnesses help the occupants decelerate at the same rate as the airplane structure. Without a seat belt, when the airplane stops, the occupant continues at the same speed as before the impact. It's the secondary stop, without restraint, that causes injuries.

Use whatever padding is available to cushion the stop.

### Speed and Stopping Distance

The severity of deceleration is governed by the square of the ratio of the actual speed divided by the desired speed. If the ground speed is doubled, the destructive energy is quadrupled. Slow down as much as practical (not possible).

An engine failure at night in a single-engine airplane is more excitement than most pilots hope for during a lifetime of flying. Fly into the wind at the slowest possible airspeed.

# CHAPTER 7

# Descent and Traffic Pattern

Any indication of turbulence during the descent to the traffic pattern should alert the pilot to maintain the maneuvering speed or turbulent air penetration speed (if applicable).

Use caution when making a rapid descent. One must be careful to avoid the two situations that cause harmful engine damage – thermal shock and detuning the engine.

## THERMAL SHOCK

If the engine is cooled too fast, thermal stresses occur in the engine. The valves may become warped and certain parts may crack or become distorted. This occurs because the engine uses different types of metals in its various components. Metals with different rates of contraction and expansion can raise havoc when cooled too fast.

For example, a turbocharger may experience temperatures of up to 1,600-degrees Fahrenheit in normal cruise flight. If the throttle is reduced to idle while descending, the internal turbocharger components' temperature may cool below freezing in a very short time. The situation becomes worse in winter.

During descent in an airplane with a constant-speed propeller, maintain the manifold pressure no lower than the bottom of the green arc, around 15 inches. For an aircraft equipped with a

fixed-pitch propeller, keep 1,500 rpm as a minimum. This keeps the engine "ready for action" if a go-around maneuver becomes necessary and prevents thermal shock.

Sometimes losing altitude is difficult while carrying power. Slow to the flaps extended speed and add flaps to create drag. Then the "extra" power required to pull the airplane through the air during the descent will keep the engine warm. If the airplane has retractable gear, lower the gear (observe the maximum gear operating and extended speeds) to create additional drag.

One may get away with thermally shocking the engine and making a go-around with full power, but the cumulative effects of the induced stress may be such that the engine will quit some day during normal operations.

## DETUNING THE ENGINE

Information about detuning the engine is difficult to find. Avco Lycoming Division (now Textron Lycoming) published Service Bulletin No. 245B to address the problem in detail. It found engine damage can occur due to detuning the dynamic counterweight system of some high-performance engines.

To explain the operation of the dynamic counterweight system, having a basic understanding of inertia and pressure forces is necessary. Inertia loads are the forces of the engine that increase with engine speed. Cylinder pressure loads are compression forces acting on the pistons that are proportional to manifold pressure. Inertia loads and cylinder pressure loads oppose each other resulting in an exciting force that causes torsional vibration in the propeller-crankshaft systems.

The dynamic counterweight system consists of floating weights on the crankshaft that are designed to negate or minimize the torsional vibration.

Detuning of the dynamic counterweight system occurs when the engine operates outside the normal range. At low speed (idle and taxi) the inertia forces are low along with the cylinder pressure, producing a small differential in forces. At high speed the inertia forces are higher, and the cylinder pressures are also high, resulting in a small differential.

When this differential is increased markedly, the amplitude of the counterweight swing may be increased beyond its travel limits, causing bumping and rattling of the counterweights against the crankshaft. Along with the rapid and severe damage to the counterweights, rollers they ride on and their bushings, this detuning negates the effect of the counterweights. It sets up severe torsional fatigue stresses in the propeller-crankshaft system that will eventually result in a major structural failure in some components of the engine.

The pilot should be aware of the four operating conditions that can cause detuning of the counterweight system.

### Rapid throttle operation

Rapid opening or closing of the throttle might cause counterweight detuning. Detuning can occur if the power is suddenly cut off, such as during a simulated engine failure in single-engine or twin-engine airplanes. To avoid this, use the mixture control, leaving the throttle in the normal open position until the engine has slowed because of lack of fuel, then close the throttle to the idle position.

The procedure of leaving the throttle in the normal open position allows the cylinder to draw in atmospheric pressure to maintain compression forces that are sufficient to cushion the deceleration.

### Propeller Feathering

When practicing engine shutdown and restart in multiengine airplanes, be sure the throttle is set approximately to zero thrust before opening the mixture control and starting the engine. With the thrust set to a higher or lower power setting the engine and propeller system will surge during the start. Use the POH or aircraft manual for specific feathering and restart instructions.

### High Engine Speed and Low Manifold Pressure

Operations involving high engine speed and low manifold pressure, such as during a power-off descent, can cause detuning of the counterweight system.

### Excessive Speed and Power

All supercharged and turbocharged engines have the inherent capability of operating at power settings beyond the capabilities of the engines, especially at low altitudes and during cold weather.

## CARBURETOR ICE

Liquid fuel will not burn in an engine. The carburetor has been designed to convert the liquid gasoline to a vapor. Vaporization of fuel in the carburetor and expansion of the fuel/air mixture as it passes through the engine to the cylinders will lower the carburetor air temperature as much as 60°F. When the air contains water vapor, ice forms in the carburetor when the temperature reaches the freezing range.

### Categories of Carburetor Icing

There are three categories of carburetor icing: fuel ice, impact ice and throttle ice.

### Fuel Ice (Fuel Evaporation Ice, Fuel Vaporization Ice)

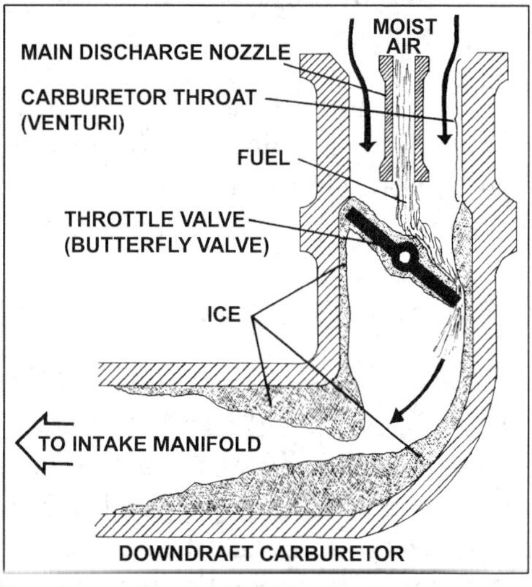


*Figure 1.* **Carburetor icing.**

Fuel ice forms downstream from the point where fuel is introduced, between the carburetor and intake manifold. It is most prevalent in airplane engines equipped with conventional float-type carburetors. Fuel ice may occur at temperatures from 32°F to as high as 100°F with a relative humidity of 50 percent or greater. Usually, potential icing conditions occur when the temperature-dew point spread is 20°F or less and the relative humidity is 50 percent or greater.

### Impact Ice (Atmospheric Ice)

This ice is formed by the impact of water (rain, snow, sleet or cloud droplets), when in suspension in the atmosphere, contacting parts of the induction system at temperatures of freezing or below. Airplanes having a carburetor, but no alternate air, are susceptible to having ice cover the air filter, choking off the air to the engine. Carburetor heat must be used ... not to melt the ice, it will not, but to provide an alternate source of air (unfiltered air) for the engine.

### Throttle Ice (Expansion Ice)

When condensed water vapor freezes near the throttle valve (butterfly valve) in the carburetor, it is called throttle ice. The effect of throttle icing is a progressive decline in power.

## CARBURETOR HEAT

Generally cruise power settings generate sufficient heat to prevent carburetor icing. When operating at reduced throttle settings, the butterfly valve closes off part of the carburetor throat and a small amount of ice could stop the engine.

Carburetor heat is an anti-icing device (not a deicing device) that preheats the air before it reaches the carburetor. The carb heat is usually adequate to prevent icing, but it will not always clear out ice that has already formed.

If the carb heat is applied too late to melt the ice, a few pilots swear you can turn off the magnetos momentarily, causing an excess of fuel in the intake manifold. When the mags are turned on, a backfire occurs that may dislodge the ice. Use this technique only during an emergency, as it has the potential of blowing the cylinder head gasket, along with other harmful effects.

Application of carb heat shunts the normally filtered air to the engine. Carb heat uses unfiltered intake air that is passed through a baffle around the exhaust manifold or the muffler. This heated air enters the carburetor throat at temperatures approaching

200°F. To be fully effective, the heat should be applied before the throttle is closed to warm the carburetor and intake system.

Carburetor heat reduces engine power because the heated air is less dense. The mixture may need to be leaned for smooth engine operation.

Using partial carburetor heat for ice prevention without a carburetor air temperature gauge may be worse than using none at all. Partial heat may raise the mixture temperature to a danger range where throttle ice forms.

## DESCENT RULE

Many pilots, on their first solo cross-country flight, arrive at the destination airport too high to land. After circling to lose altitude, they become determined to develop some system to help them decide when to begin the descent to the traffic pattern.

Some pilots use a computer to find the time necessary to descend, the number of feet from cruise altitude to the pattern altitude, and from this they calculate the required rate of descent. Others use DME, LORAN or GPS and ground speed. Some compute the ground speed versus the ETA to arrive at a distance out to begin the descent.

Whatever method works best for you is the one to use. Another method that compensates for changes in ground speed during descent involves simple mental computations rather than computer operations.

Knowing two terms is necessary to use this descent rule:

- *Altitude refers to the altitude in thousands of feet AGL. This is the altitude divided by 1,000. For example, 10,000 feet is the number 10. Or, 14,500 feet is the number 14.5.*
- *Rate of descent is equal to the ground speed divided by two and multiplied by 10. That is, take half the ground speed value and add a zero.*

The descent rule then becomes simple. Multiply three times the altitude (cruise altitude minus pattern altitude). This provides the number of miles (nautical) from the airport to begin the descent.

The rate of descent is half the ground speed divided by two.

- *3 x ALTITUDE = DESCENT POINT*
- *GROUNDSPEED/2 x 10 = RATE OF DESCENT*

This descent rule will result in an approximate three-degree glide path for the descent. It works well for single-engine, twin-engine, reciprocating and turboprop airplanes.

Example: Flying a Cessna 180 with a groundspeed of 120 knots at 12,500 feet to the Jackson, Wyo. airport with an elevation of 6,400 feet. The pattern altitude is 7,200 feet. How far from the airport should one begin the descent? Subtract the pattern altitude, 7.2, from the cruise altitude, 12.5 (remember the altitude is divided by 1,000), for 5.3. Round this to five. The distance out to begin the descent is 3 x altitude (5) or 15 miles.

The rate of descent is the groundspeed divided by two and multiplied by 10. The groundspeed is 120 divided by 2 = 60 x 10 = 600. A rate of descent of 600 fpm will approximate a three-degree glide path. Suppose the groundspeed increases to 140 knots during the descent. Then, 140/2=70x10=700 fpm. Adjusting the rate of descent is easy.

Checking the progress of the descent by multiplying the distance from the airport by three is desirable. This provides the altitude in hundreds of feet above the pattern altitude where the airplane should be. For example, when 20 miles from the destination, 3 x 20 = 60. One should be 6,000 feet above the pattern altitude. If not, adjust the rate of descent accordingly.

The descent rule is useful for crossing restrictions assigned by ATC. Suppose while cruising at 15,000 feet, ATC directs you to cross Podunk intersection at 7,000 feet. Subtract the altitudes (15-7=8). Multiply three times the altitude (3 x 8 = 24) for the miles out to begin the descent. Remember the last 1,000 feet of descent should be at a rate between 500 and 1,500-fpm. Depending on groundspeed, one can add an additional five or 10 miles to compensate for this without becoming involved with mathematical computations.

## PRE-LANDING CHECK

Establish the habit pattern of using a checklist for all phases of flight. Complete the pre-landing check on the downwind leg opposite the point of landing. This leads to consistency that can become important when distractions arise. For example, a pilot flying a retractable-gear airplane performs the pre-landing check downwind opposite the landing point. On short final, he is directed by ATC to make a go-around, and retracts the landing gear and flaps. Downwind opposite the landing point he goes through the before landing check and discovers the landing gear is not down and locked.

It is during non-standard maneuvers, procedures or distractions that the pilot forgets to extend the gear. With established habit patterns, forgetting the gear is less likely.

## NORMAL APPROACH

Because of the variables, there is really no such thing as a normal approach. Variations occur in the airplane's lateral distance from the runway on the downwind leg. The altitude to begin the descent and turn on the base leg and the final approach are seldom the same. The wind direction and velocity, the amount of flaps used, and the approach speed also alter the 'normal' approach. To establish consistency, make a stabilized approach.

## STABILIZED APPROACH

The secret to a good landing is a good approach. The secret to a good approach is consistency. To make the approach become routine, fly the traffic pattern at the same altitude above the ground each time. Fly the same distance from the runway each time. If the distance from the runway of one-quarter mile is desired, determine where on the strut (high-wing airplane) or the wing (low-wing airplane) the runway intersects. Then fly this position each time. Do the pre-landing checklist opposite the point of landing each time. Turn base when the aiming point (the point on the runway where the flare is initiated) is half way between the wing tip and tail, each time.

# CHAPTER 8

# Landing

## CAUTION

A word of caution for the neophyte taildragger pilot should be expressed. It is important to maintain your attention to the phase of flight being accomplished. Students, especially during crosswind operations, have run off the runway after landing while messing with the flaps. They are very attentive during the approach and landing, but they have been told that retracting the flaps to place more weight on the main wheels is important. Usually the runway excursion occurs when they try to stow the Johnson bar (a mechanical flap handle found between the seats of some airplanes). If the flap handle cannot be fully retracted and locked without looking down or without blocking your vision, retract the flaps as far as possible while directing your attention outside, and then postpone full-flap retraction until after coming to a stop.

## INSTRUCTORS

A taildragger instructor, observing inattention in a student, often admonishes the student to direct his attention to the particular phase of flight (usually during the landing roll-out). Later, preoccupation or indifference may cause the student's mind to again wander away from the task at hand. If this happens during the after-landing roll-out, allowing the student to run off the runway once—if it can be done safely—may stop this inattention from ever happening again. This is one way to reinforce the admonition that, "You always fly the darn thing until it's tied down."

---

**Flying is the second greatest thrill known to man ... Landing is the first.** – *Aviation cliché*

---

Most of the mistakes a student encounters during training should be done under the supervision of the instructor rather than allowing the student to experience them on his own. The inventive instructor, when his student does not make errors on his own, will create scenarios that will expose the student to the 'problems' facing taildragger pilots.

The pilot must learn: "It's not your mistakes that will get you; it's how you take care of them." Be sure to impart the knowledge of mistakes—inattention, the bounce, the porpoise, and go-around maneuvering—and how to correct for them.

## STABILIZED APPROACH

A stabilized approach for landing is used for all landings—normal, soft-field and short-field. Prior to about 1968, instructors taught the normal landing as, "Downwind opposite the landing point, reduce throttle to idle." The student then judged the glide to arrive at the runway. Turbulence, wind shear, head wind, or a lack of judgment often necessitated a go around. At times the engine did not respond to the throttle command and go-around accidents resulted. This caused a re-thinking of the approach technique and the advent of the 'stabilized approach.'

The secret to consistently good landings is a good approach. The stabilized approach is a good approach.

So what is a stabilized approach? An approach where the pilot has a consistent airplane configuration, maintains an exact vertical flight path on final to a predetermined aiming point on the runway, and maintains an exact airspeed while flying the glide path. In other words, the stabilized landing approach means flying an exact glide path without an electronic aid, while maintaining an exact airspeed.

Power is maintained during the approach to keep the engine warm and ready to respond when needed. Also if carburetor heat is used, partial power guarantees greater effect than idle power. The power setting should result in a rate of descent appropriate to the desired glide path angle. This will be approximately 350 to 650 fpm.

> *(See: Spot Method For Landing – Best Approach Angle* to determine the approach angle and rate of descent for your particular airplane*)*.

The power setting to achieve the desired glide path angle may be about 1,500 rpm for the fixed-pitch propeller airplane and around 12-inches manifold pressure for the constant-speed propeller airplane.

## Procedure

When approaching an airport, make gradual power reductions. The initial power reduction may be to 20-25 inches (or 2,000 rpm with a fixed-pitch propeller). After several minutes (one to three minutes), reduce the power further to 17-20 inches (1,800 rpm). In the pattern, reduce to about 12-17 inches (1,500 rpm). This gradual reduction of power allows the engine temperatures to cool slowly and equalize. The different metals with different rates of contraction need a little time to adjust.

---

> **My airplane is quiet, and for a moment still an alien, still a stranger to the ground, I am home.**
> *– Richard Bach, 'Stranger to the Ground,' 1963*

---

Using this technique will add to the longevity of the engine while providing the benefit of an engine that is ready for either application of go-around power or a total reduction to idle for the landing.

Obviously the stabilized approach protects the engine. Additionally, it makes the approach for landing a safer operation since it results in a consistent performance that provides the pilot with a reduced workload and the time to make assessments of other variables. Performing the approach with the same configuration, descent rate and airspeed increases the pilot's situational awareness of a headwind (rate of descent needs to be less) or tailwind (rate of descent needs to be more) during the approach.

## EFFECT OF FLAPS

Extracting maximum performance from an airplane is something each pilot strives to do no matter what airplane he flies. Flap management is an important factor in obtaining that performance.

---

**You don't have to be in *Who's Who* to know what's what.** — *Anonymous*

---

The purpose of flaps is to allow a steep approach angle to the runway without an increase in airspeed. Flaps change the camber of the wing, increasing the lift and the drag.

Lift and drag are directly proportional. If lift is increased, drag is increased. The addition of the first 50 percent of flaps causes more lift than drag (in most airplanes). The addition of the last 50 percent of the flaps causes more drag than lift.

During the execution of a go-around, the flaps should initially be retracted to 50-25 percent (half to one-quarter). The remaining flaps should not be retracted until the airplane has sufficient speed to sustain flight without sinking or stalling.

Flap management requires prior thought on the consequence of flap extension and retraction. Extending the flaps causes the following changes to occur:

- *Lowering the flaps changes the camber of the wing, requiring re-trimming to balance the nose-down moment change.*
- *Lift and drag are directly proportional. Increase lift and drag increases. The increase in drag*

*requires a higher power setting to maintain a constant airspeed at a constant altitude.*

- *The angle of attack required to produce the same lift coefficient is less with the addition of flaps and will cause the airplane to balloon.*
- *Excessive airspeed—above $V_{FE}$—when lowering flaps for landing, or exceeding the airspeed limitation once the flaps are extended, may cause structural damage of the wing.*
- *The addition of flaps reduces the structural strength of the wing. Various makes and models may have the g-force tolerance, with flaps extended, reduced 8 to 58 percent.*

If the flaps are retracted prematurely (insufficient airspeed), the lift coefficient of the "clean" wing probably cannot support the airplane.

In making the transition from "dirty" to "clean" configuration, three changes take place:

- *The reduction in camber by flap retraction changes the wing pitching moment—for most airplanes—and requires re-trimming to balance the nose-up moment.*
- *The retraction of flaps causes a reduction in drag that improves the acceleration of the airplane.*
- *The retraction of flaps requires an increase in the angle of attack to maintain the same lift coefficient. Thus, if airplane acceleration is slow through the flap retraction speed range, lower the nose to increase airspeed before flap retraction to prevent sinking. Consider "milking" the flaps, that is, retract them slowly, a bit at a time.*

## Flaps for Landing

The FAA created a controversy many years ago when they advocated the use of full flaps for all landings, even in crosswinds.

They did their best to educate pilots about the benefits of full-flap landings in crosswinds.

Some pilots feel a no-flap landing is easier to execute and more controllable. The following information is presented, not in an attempt to convert anyone, but to provide a method for logical experimentation. In this way, the pilot can choose flap, partial flap or no-flap landings, based upon knowledge rather than myth.

To learn the degree of controllability of any particular airplane with and without flaps, try this experiment. At some speed less than the maximum flap operating speed, roll the airplane into a 20- to a 30-degree bank. Due to the inherent stability of the plane, it will tend to roll back to wings-level flight without maintaining aileron pressure. While holding the bank constant, extend one-quarter flaps, then try half-flap and full-flap extension. If the bank becomes noticeable steeper, it suggests the addition of flaps increases the aileron's ability to control the airplane about the roll axis.

For some airplanes this is true. The airplane may have more "air control" with the addition of flaps. It is also true the airplane has less "ground control." The flaps, being behind the main gear, allow any crosswind to create more weather vane tendency once the airplane is in contact with the runway. This is one reason why some pilots are hesitant to use flaps during crosswind operations.

Flaps can also direct the airflow away from the elevator and rudder, making them less effective in crosswind landings.

---

**That's not flying, that's just falling with style.**
*– Woody, from the 1996 movie 'Toy Story,' regarding Buzz Lightyear.*

---

If flaps are used, proper technique calls for retracting the flaps once on the runway, then continue to fly the airplane on the ground by making crosswind corrections with the ailerons.

Some airplanes afford more aileron control with the flaps extended. If the flaps do not have a detrimental effect on the elevator and rudder, the use of flaps may provide for a safer landing in crosswinds. This is due to the decrease in centrifugal force if a swerve is encountered. Centrifugal force increases as the square of the speed where it starts.

Retract the flaps—not the gear—immediately upon becoming established as a ground vehicle. This improves braking ability by placing more weight on the wheels, improves tail wheel steering by allowing more weight on the tail wheel, and reduces the natural weather vane tendency. If full flap retraction requires momentarily being unable to see the runway, retract the flaps partially and wait until the airplane is stopped to complete the retraction. It's better to deal with the control problems of partial flaps than it is to experience a swerve because you haven't maintained sight of the runway.

When the airplane starts to weather vane, the mass located behind the c.g. wants to continue straight along the original path. When the pivot point (main wheels) move away from the straight-line ground path, the tail will start to move opposite the main gear and a swerve is created. Suppose one has a choice of a touchdown at 70 KIAS without flaps and 50 KIAS with flaps. Seventy squared is 4,900, while 50 squared is 2,500. So any swerve encountered at 70 KIAS would be nearly twice as strong as at 50 KIAS.

### Caution — *"Avoid Slips With Flaps Extended"*

If an airplane is placarded against slips with flaps extended, it is because the flaps direct the airflow away from the tail. If the pilot slips such an airplane, then makes a rapid recovery to coordinated flight, the tail may stall and the nose may pitch down steeply.

For example, slipping to the left causes the right horizontal stabilizer/elevator to be blanked. A rapid recovery causes the left horizontal stabilizer/elevator to be blanked before the right can regain airflow. With the normal negative lift of the tail removed because of the lack of airflow (tail stall) the nose pitches down rapidly.

The author has tried to demonstrate this tail stall to students over the years. Surprisingly enough, out of the hundreds of attempts only two have resulted in the actual tail stall with the rapid nose-down pitching moment. The weight and balance was approximately the same each time, so the c.g. location wasn't the reason for the difference.

This experimentation that has resulted in the majority of slips not causing any dire circumstance has not provided enough confidence to disregard the placard warning—*Avoid Slips With Flaps Extended*—as it only takes one time when the tail actually stalls close to the ground to ruin your life.

## LANDING DEFINITIONS

With the exception of crosswind landings that are covered in Chapter 10, this section will discuss normal landings—three-point and wheel, soft-field landings, short-field landings, landing irregularities and the simplified landing.

---

**There is an art ... to flying. The knack lies in learning how to throw yourself at the ground and miss.** — *Douglas Adams, 'The Hitchhikers Guide to the Galaxy'*

---

Prior to the discussion of the various landings, a few definitions are in order so we compare apples to apples.

### Downwind Leg

Practicing conventional-gear landings will be easier if the same thing happens at the same time during each approach. The secret is to become consistent. Becoming consistent in flying the downwind leg is important in making a good approach. Consistency means flying the same distance from the runway at the same altitude each time.

### Base Leg

Using a stabilized approach, that is, carrying partial power that results in an approximate 500-fpm rate of descent, allows the turn from the downwind leg to the base leg to be made with consistency.

On the downwind leg, when the aiming point on the runway (the point where the flare is initiated) is midway between the wing tip and the tail, turn onto base. This also leads to consistency.

When the turn from base to final is accomplished with the airplane aligned with the extended centerline of the runway, it is easy to detect wind drift.

## Final Approach

Use the same indicated airspeed for a normal approach to landing regardless of the physical or density altitude. We are aware that at high-elevation airports the air is thin. This thin air reduces the lift of the wings, the power output of the engine and the thrust of the propeller. But, the same thin air that affects the aircraft performance also affects the airspeed indicator.

Assuming no instrument or position error of the airspeed indicator, an airplane operated at sea level under standard conditions will indicate the true airspeed, suppose 70 KIAS. At 5,000-feet density altitude the air is less dense. By the time sufficient air, by weight, registers 70 KIAS, the airplane is actually going faster. Remember the rule of thumb that the true airspeed is approximately two percent faster than indicated airspeed for each thousand feet above sea level. This creates a built-in compensating factor that allows the use of the same indicated airspeed for approach to landing regardless of altitude.

### Rule of Thumb - Determine True Airspeed

*The true airspeed is approximately two percent faster than indicated airspeed for each 1,000 feet above sea level.*

Although the same indicated airspeed is used and a stabilized approach is used, it will be necessary to use a little more power during the stabilized approach to high-elevation airstrips to have the same rate of descent that is observed at sea-level strips.

---

**There are three simple steps for making consistently smooth landings: First is the approach; second is airspeed control; and unfortunately, no one knows the third.** — *Anonymous*

---

Vision is important during the final approach and flare. Let the head assume a normal position. Rather than moving your head back and forth as the nose comes up in the flare (assuming a three-point landing), use your peripheral vision by moving your eyes instead of your head.

When focusing on one spot on the ground, it is difficult to develop an altitude perspective. By slowly and constantly changing the focus from side to side and from the aircraft nose to the horizon, the brain, without one realizing it, chooses a number of points for comparisons. To change focus, move the eyes, not the head. This technique makes it possible to more accurately judge the height and movement of the airplane.

The approach speed of 1.3 $V_{SO}$ is the speed used for maneuvering during the approach for normal landings. This speed allows a margin of 30 percent above the stall speed to compensate for maneuvering. The speed should be reduced to about 1.2 $V_{SO}$ for the over-the-fence approach.

'Over-the-fence' is an expression used to explain the position on final when the aircraft crosses the runway threshold at approximately 20- to 30-feet agl and in a position to make a normal landing.

---

**To put your life in danger from time to time... breeds a saneness in dealing with day-to-day trivialities.** – *Nevil Shute, 'Slide Rule: The Autobiography of an Engineer'*

---

## Flare

If the pilot uses the airplane attitude for flying, the flare is easy to accomplish. Attitude is the position of the nose (and the wings) in relation to the natural horizon.

Ideally, the flare should begin at 10 to 15 feet above the runway. There is a smooth transition from the glide attitude to the level-flight attitude. With reduced power and level-flight attitude, the airplane will begin to settle. As the airplane settles, begin a slow transition from the level-flight attitude to the landing attitude. (Pictures of these attitudes is shown later under *Simplified Landing.*)

When the glide attitude (the position of the nose below the horizon) is maintained the airplane will provide a constant performance. That is, a specific airspeed results from maintaining the glide attitude.

*A constant attitude and a constant power setting results in a constant performance. This is true regardless of the phase of flight such as climbing, cruising or gliding.*

While maintaining the glide attitude the airspeed indicator can be covered and with a minimum amount of practice to find the 'sweet spot,' the pilot can fly an exact airspeed, usually within one knot of that desired.

The level-flight attitude is the position of the nose below the horizon during level flight at cruise airspeed and cruise power setting. With practice it is easy to determine, remember and simulate this attitude.

The landing attitude approximates the normal climb attitude. During climb the nose will be above the horizon. Remember where the horizon intersects the side of the nose cowling for transition to the landing attitude.

## Touchdown

The approach and flare are the same for a three-point landing or a wheel landing.

For a three-point landing the flare is continued to the landing attitude, that is, the attitude that results in the main wheels and the tail wheel all touching the runway surface at the same time.

*Figure 1.* **A landing or an arrival?**

The wheel landing is different only in the fact that after the flare, partial power keeps the airplane in level flight until the wheels touch the surface. A wheel landing may be accomplished from the tail-low landing attitude also.

Occasionally, when the approach speed is too fast, the airplane will float along. The wheels are only a couple of inches above the ground, but the airplane doesn't want to land. Making the airplane land provides moments of excitement or apprehension.

*A trick that works well for all landings in tail draggers (and airplanes with training wheels), singles and twins, is to roll the aircraft slightly to one side or the other with a maximum wing-tip deflection of about six inches from the level position. One wheel will 'kiss' the ground and the other will follow. This will not work if the airplane is more that several inches above the runway.*

### After-landing Roll

After the touchdown, it is important to use the ailerons to maintain a wings-level attitude. The nervous student, in a crosswind, often applies full aileron toward the wind once he is on the ground. This increases the ground loop tendency by rolling the plane into the wind and creating a downward force on the upwind tire that leads to extra drag.

Feel what the airplane is doing. Use the ailerons to maintain the wings level. As the airplane slows, the ailerons become less effective. The pilot must use more and more aileron deflection as the speed decreases.

Once the tail is on the ground the elevator control should be "sucked into your gut," that is, it is held back firmly as far as it will go. This places weight on the tail wheel and provides more steering authority. If the airplane touched down in the three-point attitude, moving the elevator control full aft will prevent bouncing or skipping.

---

**Now I know what a dog feels like watching TV.**
*— A DC-9 captain trainee attempting to check out on the 'glass cockpit' A-320*

---

## GROUND EFFECT PHENOMENA

Every pilot has encountered the term "ground effect." As an expression, it is frequently tossed about in "hangar sessions." Although the term is widely used, its effect on airplane performance may be but dimly understood.

Ground effect works during takeoff and landing. In the common case of an airplane approaching with excessive speed, it flies down from the free air into ground effect. The reduction of induced drag as the airplane nears the runway comes into effect to make the airplane float. This leads to the classic type of overshooting.

Even with normal approach speeds, the dissipation of airspeed in ground effect takes longer, resulting in floating, especially during hot days. On short fields, approach as slowly as is consistent with safety to reduce the ground effect. When you overshoot and the far end of the runway is fast approaching, recognize that ground effect is helping the plane to float. Providing there is ample room, give it the gun and go-around when you realize you have approached too fast and are skating along in ground effect.

---

**No need to teach an eagle to fly.** – *Greek Proverb*

---

## Drag and Ground Effect

The total drag of an airplane is divided into two components, parasite drag and induced drag. These two are always present, though in varying degrees. Parasite drag is caused by skin friction and the turbulence of the air flowing past various parts of the airplane. Induced drag, a term tossed about casually by engineers but only dimly understood by many pilots, is a result of the wing's work of sustaining the airplane in flight.

The wing lifts the airplane simply by accelerating a mass of air downward. It is perfectly true that reduced pressure on top of an airfoil is essential to lift—Bernoulli's principle—but still, it is but one thing that contributes to the overall effect of pushing an air mass downward. Sir Isaac Newton formulated the laws of motion. The third law essentially states: "Every action has an equal and opposite reaction," (every action is opposed by an equal and opposite reaction). The more downwash there is, the harder the wing is pushing the mass of air down. At high angles of attack, induced drag is high. Since this corresponds to lower airspeeds in actual flight, it can be said that induced drag predominates at low speeds and parasite drag is greatest at high speeds.

During approach for landing, the pilot exercises proper technique to control his flight path. The old adage: Attitude plus power equals performance is still important in maintaining safety of flight. If the attitude and power remain constant, the airspeed will remain constant. If the power is changed, the attitude must change to maintain the airspeed constant. If the attitude is changed, the power must be changed to maintain the airspeed constant.

When landing, if the pilot interprets that the airplane is below the desired glide path, his first reaction must not be to ease the nose up. An increase in angle of attack without an increase in power will lower the airspeed and greatly increase the induced drag. Such a reaction will create a higher rate of descent and may lead to potentially undesirable consequences. When operating on the backside of the drag curve, the elevator (angle of attack) controls airspeed and power controls rate of descent or climb.

---

**If black boxes survive air crashes -- why don't they make the whole plane out of that stuff?**
*– George Carlin*

---

## Ground Effect

Ground effect has been explained as a "cushion of air" between the wing and the ground. It "ain't" so. Ground effect works on wing downwash and wing-tip vortices. It reduces the angle of downwash and diminishes the effect of wing-tip vortices. This results in a reduction of the airplane's induced drag when very near the ground.

Contrary to some explanations of lift, the relative wind (near the airfoils) is not parallel to the longitudinal axis of the airplane. Rather, there is an upwash ahead of the wing and a downwash behind the wing. When the wing is flown close to the ground, the upwash and downwash cannot develop. The wing can produce sufficient lift for flight at an angle of attack that is less than that required at higher altitudes—or so the wing thinks. Actually, the induced drag is reduced.

At the moment of liftoff—when the wing is at a height equal to 10 percent of its span above the ground—there is about a 48 percent reduction in the induced drag as compared to flight at altitude. Induced drag increases rapidly as the plane climbs. At a height equal to 50 percent of its span above the ground—somewhere around 18 feet—there is only about an eight percent reduction in induced drag due to ground effect.

### Extent of Ground Effect

A pilot trying to takeoff from a poor field uses full power and holds the plane in a nose-high attitude. Ground effect reduces induced drag so the airplane can reach a speed where it can be "horsed" off. However, as altitude is gained, induced drag increases while the ground effect diminishes. Seemingly clear of the ground, the airplane disastrously falls back to earth.

Twenty to 30 feet up, ground effect vanishes, the wing encounters the full effect of induced drag and the struggling plane, that got off the ground on the ragged edge of a stall, becomes fully stalled and drops to earth.

A mixture of short runways, rough ground, grass and snow, high airport elevation, high air temperature, a weak engine and a heavy load—in any of many combinations—are a danger signal. When facing a marginal takeoff situation, know the plane's takeoff speed for the conditions prevailing, the distance required to accelerate to the speed, and then allow a generous margin of safety by picking up as much speed as practical just off the ground, flying in ground effect, before trying to climb. If the plane is still dragging its wheels when it should be airborne, abort the takeoff and wait for better conditions or off-load the plane.

It is desirable to accelerate after a soft-field or short-field takeoff before climbing. To do this, maintain the wing as close to the ground as possible. While accelerating to $V_X$ or $V_Y$, fly no more than one-half the wing span above the ground. (See Chapter 5, 50/70 Rule for Takeoff)

## SPOT METHOD FOR LANDING

The spot method for landing, hereafter called the "spot method," is applicable to each and every landing, not just the short-field landing. It is particularly useful when visual illusions or sloping terrain can cause landing problems where the pilot may land short or overfly the runway.

The solution for making the perfect approach consists of:

- ✔ *Maintaining a constant airspeed on final.*
- ✔ *Pointing the nose (windscreen mark) at the aiming point (spot) on the runway.*
- ✔ *Controlling the nose position (windscreen mark) with the throttle.*

To explain the spot method requires an understanding of the approach speed and approach angle.

## Optimum Approach Speed

The optimum approach speed during stable conditions is a speed equal to 1.3 $V_{SO}$ ($V_{SO}$ is the stall speed or minimum steady flight speed in the landing configuration).

The optimum approach speed is not the "over-the-fence" speed. There is a transition from the optimum approach speed to the over-the-fence speed on short final. To find the optimum approach speed for any aircraft under varying conditions of weight and center of gravity, fly the airplane to a safe altitude. A safe altitude is one that is reasonable and prudent to practice stalls.

---

**The airplane is just a bunch of sticks and wires and cloth, a tool for learning about the sky and about what kind of person I am, when I fly. An airplane stands for freedom, for joy, for the power to understand, and to demonstrate that understanding. Those things aren't destructable.** — *Richard Bach, 'Nothing by Chance,' 1963*

---

At altitude, perform a pre-landing check, extending the flaps to the landing configuration (and the gear, if applicable), while slowly reducing the throttle to idle. It's probably a good idea to make progressive power reductions with a short period of time at each reduced power setting to stabilize the engine temperature. This prevents thermal shock. Do not de-tune the engine by making rapid throttle movements.

Maintain level flight. This will result in a slow deceleration of the airplane until a stall occurs. Note the airspeed when the stall buffet occurs, then make a normal stall recovery.

Multiply the indicated stall speed by 1.3. This is the optimum approach speed during stable conditions.

*NOTE: The optimum approach speed is not the "over-the-fence" speed used for landing. The optimum approach speed provides a margin of safety above the stall speed while maneuvering to the final approach position. On short final the airspeed is slowed to the over-the-fence speed of 1.2 $V_{SO}$.*

## Stabilized Approach

What electronic aids can you use for glide slope guidance? The ILS glide slope (instrument landing system), VASI (visual approach slope indicator), or PAR (precision approach radar, sometimes called GCA—ground controlled approach).

It is unlikely any of these aids will be available at unimproved airstrips. But, flying a predetermined flight path or glide slope is possible and desirable.

A stabilized approach is used for all landings. The stabilized-landing approach means flying an exact glide path without an electronic aid, while maintaining an exact airspeed.

**Experience is the knowledge that enables you to recognize a mistake when you make it again.**
*— Aviation Cliché*

## Best Approach Angle

A 3-degree glide slope is not necessarily the most favorable for light aircraft. The use of a steeper approach angle, up to 4.5 degrees, may make it easier to see the runway and may make it easier to correct the flight path during the approach.

## 3-degree Glide Slope

To fly a 3-degree glide slope, multiply the groundspeed in knots by five. This gives the rate of descent to maintain a 3-degree glide slope.

Flying a predetermined flight path (glide slope) results in a stabilized approach. The stabilized approach is one of the keys to a good landing.

Assume the approach speed is 80 knots in a no-wind condition, resulting in 80 knots groundspeed. What is the required rate of descent for a 3-degree approach angle? Multiply 80 x 5 = 400 feet per minute.

## 4.5-degree Glide Slope

To determine a 4.5-degree glide slope, multiply the groundspeed in knots by eight. For example, 80 x 8 = 640 fpm.

---

**Lack of planning on your part does not constitute an emergency on my part.** — *Aviation Cliché*

---

## Windscreen Mark

The windscreen mark is a point on the windshield used to aim at or align the airplane with the aiming point.

With experience you learn to visualize the windscreen mark without conscious thought and without a physical mark on the windscreen. The best method to use when beginning to develop this technique is to use a grease pencil (or tear a small portion from the sticky part of a Post-it® note) and make a mark on the windscreen in front of your field of view. The mark will be about one to four inches up from the dash, depending on your parallax.

Parallax affects the position of the windscreen mark and is an apparent change in the direction of an object, caused by a change in observational position that provides a new line of sight. If the position of the seat is moved forward or aft, or raised or lowered, it will change your parallax. This is the reason for adjusting the seat to the same position each time you go flying.

A more accurate method of determining the windscreen mark, rather than marking the windscreen one to four inches up from the dash, is to determine the mark in flight. At cruise power setting and cruise airspeed, note where the horizon intersects the windscreen. Make the windscreen mark one-half inch below this point of intersection. Try to find some distinguishing characteristic on the windscreen at this point, otherwise, when you bend forward to make the mark, the horizon will move and the mark will not be in the proper place (it will be too low).

*Figure 2.* **Level-flight attitude from cruise flight.**

Do a test to see if the selected windscreen mark is placed properly. Perform a pre-landing check and begin a descent. Look at the windscreen mark and look at the ground. The point on the ground should remain stationary, aligned with the windscreen mark. The ground below the mark will be moving down and under the nose. The ground above the mark will be moving up on the windscreen. If not, make one-half inch corrections until the windscreen mark is determined.

## Landing Requirements

For all landings, the pilot should choose the following on the downwind leg of the approach:

- *AIMING POINT*
- *TOUCHDOWN POINT*
- *GO-AROUND POINT*

## Aiming Point

The aiming point is the point that the airplane would hit the runway if you do not flare. This will normally be 100-300 feet down the runway from the approach end; however, on short-short fields, it may be 100 feet before the approach end.

## Touchdown Point

The touchdown point will be some 200-400 feet beyond the aiming point when using the proper over-the-fence speed.

## Go-around Point

The go-around point is that point on the runway where, if the airplane is not on the ground, sufficient runway does not remain to complete the landing.

Assume, for example, an airplane requires 400 feet to land. Approaching a runway with a length of 800 feet means the go-around point will be one-half the distance down the runway from the approach end. If you can touch down before that point, the airplane can be stopped safely.

## Spot-landing Technique

On the downwind leg of the landing pattern, perform a pre-landing cockpit check. Adjust the airspeed to the optimum approach speed and adjust the power to obtain the desired rate of descent for the best approach angle.

*Figure 3.* **The windscreen mark (offset to the right) is aligned with the aiming point on the runway.**

On final approach, place the windscreen mark on the aiming point and trim the airplane to maintain a constant indicated-approach speed. The flare is initiated at the aiming point, with touchdown 200 or 400 feet beyond. This technique makes it possible to land the airplane exactly where you determined on the downwind leg.

## The Pitch/Power Controversy

Does the pitch control altitude or airspeed? And what about the power control? This subject is nearly as controversial as making a statement that you should always perform a wheel landing during crosswind operations.

We essentially deal with *kinetic energy*—the energy of motion, *potential energy*—the energy stored as altitude, and *chemical energy*—the power of the engine for the operation of the airplane.

Try the following technique and see if it works for you ... pitch to airspeed and power to the glide path.

When the airplane is properly trimmed for an approach to landing, pitch changes are used to maintain the airspeed. If you are too low on an approach the addition of power does not accelerate the airplane, it causes an upward pitch change that makes the airplane climb, and thus realigns the airplane on the proper glide path.

A constant power and a constant attitude result in a constant performance, that is, a constant airspeed. If the attitude and power remain the same during a no-wind approach, the rate of descent will remain constant.

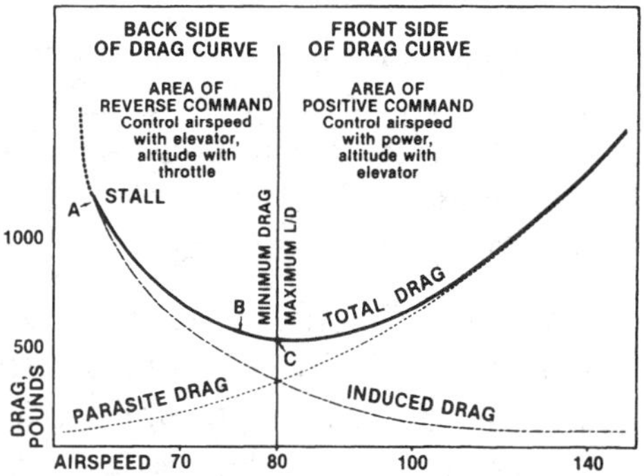


*Figure 4.* **The Drag Curve. When operating on the back side of the drag curve, control airspeed with elevator and altitude with throttle.**

Changing the power to maintain the windscreen mark on the aiming point requires a change in pitch attitude to maintain a constant approach speed. For the properly trimmed airplane, this, for the most part, occurs automatically.

For example, if the windscreen mark is aligned with the aiming point, and the aiming point begins to move down on the windscreen, the airplane is becoming too high. Reduce power and the nose will drop by itself to realign the windscreen mark and the aiming point.

If the aiming point moves up on the windscreen, the airplane is too low. Add power and the nose will rise, maintaining a constant airspeed and the accompanying decrease in the rate of descent as the nose realigns the windscreen mark and the aiming point.

## Pitch Changes

An airplane that has been properly trimmed to maintain a constant airspeed and constant rate of descent, results in "automatic" pitch changes when the power is changed.

Because of the inherent stability of the airplane, the addition of power will result in the nose rising by itself and the indicated airspeed remains constant. A reduction of power will cause the nose to lower by itself, again maintaining a constant approach speed.

Some airplanes will require small trim changes to maintain a constant indicated airspeed.

### Wind Shear Awareness

While on final approach for landing a change of groundspeed shows either a head wind or tail wind. Since the probability of having DME to give a groundspeed read is remote (although GPS is becoming more common), there is another means of determining when the groundspeed changes.

Suppose while on a 3-degree glide slope, that requires 400 fpm rate of descent with a ground speed of 80 knots, you notice the rate of descent is 500 fpm while maintaining the windscreen mark on the aiming point. What does this mean? The airplane is experiencing a tail wind. The groundspeed has increased to 100 knots. This is actually easy to calculate without delving into complicated mathematics. Recall that groundspeed multiplied by five results in a 3-degree glide slope, then 5 times “x” = 500, “x” = 100 (knots).

If you notice the rate of descent is 300 fpm, there is a head wind. Five times “x” = 300, “x” = 60 (knots). The groundspeed has decreased to 60 knots.

### Spot Landing Uses

The spot-landing technique results in accurate landings. Well, we need a qualifier here. The spot-landing technique results in the arrival of the airplane at a pre-selected point above the runway. Providing proper approach speed is maintained, the airplane will touch down within 100-200 feet of this point. Practice this technique until it becomes automatic for all landing approaches, not just at short, unimproved strips.

Once the spot method is mastered, it becomes useful for other aspects of flight:

- *Descent to pattern altitude*
- *Forced landings*

- *Crossing ridges*
- *Collision avoidance*
- *Cloud avoidance*
- *Landing uphill*
- *Landing downhill*
- *Landing in rain*

## Descent to Pattern Altitude

To reduce the risk of a midair collision, the airplane should enter the traffic pattern at the designated pattern altitude.

Select an aiming point on the ground while approaching to enter the traffic pattern. The aiming point will be the point where it is desired to enter the pattern. Align the windscreen mark with the aiming point and level off at pattern altitude.

## Crossing Ridges

The visual aspects of flying in mountainous terrain can be deceiving. During level flight, the windscreen mark is used to decide the aircraft's altitude in relation to a ridge. If the ridge line (level flight) is below the windscreen mark, the airplane is above the ridge.

## Forced Landings

With an engine failure, the priority of the pilot is to fly the airplane. Establish the best glide speed. Look at the windscreen mark to learn the aiming point on the ground. This is how far you can glide in a no-wind condition. If a suitable landing area does not exist straight ahead, mentally subscribe an arc from the aiming point. This defines the area in that you can choose a suitable landing area. The airplane can glide to any point within the area defined by the arc.

## Collision Avoidance

To determine a collision course, use of the windscreen mark is not necessary, but the principle is the same. If an airplane appears on any window, and that airplane remains in a constant spot with converging flight paths, you are on a collision course. If the airplane moves from the spot, up or down, back or forth, there is no collision course.

## Cloud Avoidance

To avoid clouds during a climb or a descent, look at the windscreen mark in relation to the cloud. If the cloud moves from the mark, you will not hit the cloud.

VFR pilots occasionally climb to get above a cloud layer and proceed toward their destination. Upon arrival they find the forecast scattered or broken condition does not exist. After flying around, they eventually find a hole. But, it is a small hole that is difficult to get through. If the pilot has no technique for getting through the hole, it can be a frightening experience at best. The preferred technique is to select an aiming point. Align the windscreen mark on the aiming point and the clouds will be avoided.

## Landing Uphill

A visual illusion occurs when landing on upslope terrain. This illusion makes it appear that the nose has a very low attitude and that the altitude is too high. To eliminate this visual illusion, use the "spot method" for the approach and landing.

## Landing Downhill

Landing downhill produces the illusion of being too low. Consequently, the pilot may approach too high and will find there is insufficient runway remaining to land. The "spot method" eliminates this visual illusion.

## Landing in Rain

Rain produces a visual illusion of appearing to be higher than the actual altitude. Use the "spot method" when landing in rain.

## Summary

The spot method for landing is a technique useful for making accuracy landings, for eliminating undershooting on upslope runways, and for eliminating overshooting on downslope runways (as explained earlier).

- *Make a windscreen mark with a grease pencil about two to four inches from the base of the windscreen, directly in front of your line of vision.*

- *On downwind, perform a pre-landing check. Reduce power to obtain the desired approach angle. When the airspeed is stabilized at the optimum approach speed, trim the airplane to maintain this speed.*
- *Determine the aiming point. Choosing the end of the runway will result in touchdown about 100 to 300 feet beyond the aiming point, depending on approach airspeed. For short runways selecting the aiming point approximately 100 feet before the end of the runway may be desirable.*
- *When the windscreen mark is aligned with the aiming point, the airplane is on the correct glide path.*
- *If the aiming point is below the windscreen mark, the airplane is high. Reduce power and check the alignment of the windscreen mark and the aiming point.*
- *If the aiming point is above the windscreen mark, the airplane is low. Increase power until the windscreen mark and aiming point are aligned.*
- *Transition to 1.2 $V_{SO}$ for the over-the-fence speed. If the airplane does not touch down within 200-300 feet after the flare, transition to 1.15 $V_{SO}$.*

## LANDING AERODYNAMICS

### Effect of CG on Directional Control

The longitudinal axis of the airplane must parallel the movement of the airplane at the moment of touchdown, otherwise a moment force is produced about the pivot point, the main wheels.

This is not so much different from the 'moment of truth'—*a portion of time during a crisis on whose outcome much or everything depends*. But in this case 'moment' is defined as a tendency to produce motion about a point.

The difference between the direction the airplane is traveling and the direction the airplane is pointed produces the moment force that causes a swerve when the wheels pivot to allow the airplane mass to parallel the airplane motion due to inertia.

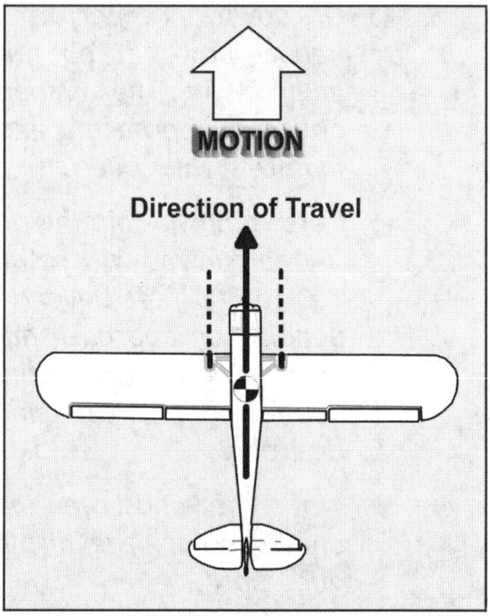


*Figure 5.* **When the inertia of the direction of travel parallels the direction of motion there is no swerve tendency during the touchdown.**

Inertia is the property by which the mass of the airplane remains in uniform motion in the same straight line unless acted upon by some external force.

Newton's First Law of Motion states, "...an object in motion tends to stay in motion with the same speed and in the same direction unless acted upon by an unbalanced force." In other words, objects tend to keep doing what they're doing. It is the natural tendency of an object to resist changes in their state of motion. This tendency is described as inertia.

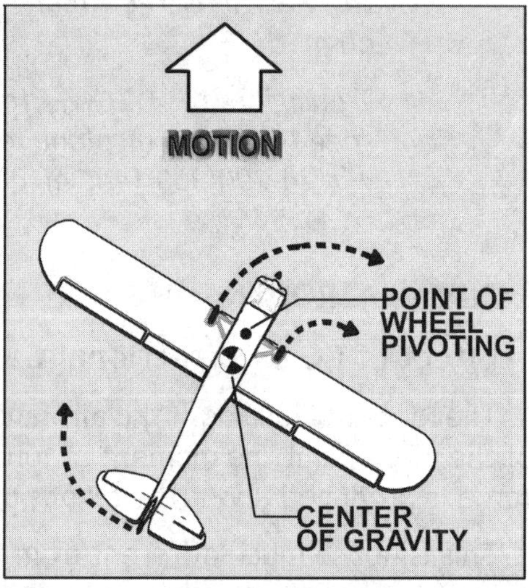


*Figure 6.* **When the airplane touches down at an angle to the motion a swerve occurs.**

Newton's Second Law of Motion asserts, "The acceleration of an object as produced by a net force is directly proportional to the magnitude of the net force, in the same direction as the net force, and inversely proportional to the mass of the object." This provides the explanation for the behavior of an airplane that does not touch down straight. The forces do not balance. The law states that unbalanced forces cause objects to accelerate with an acceleration which is directional to the net force and inversely proportional to the mass. This explains the reason for being unable to contain the airplane once a large swerve is allowed to occur. The controls (rudder, elevator and ailerons) do not have the 'power' to overcome the centrifugal force generated.

Newton's Third Law declares, "For every action, there is an equal and opposite reaction." This means that in each interaction there is a pair of equal and opposite action-reaction forces.

Contemplate the movement of the airplane with misalignment of the longitudinal axis in relation to the overall motion. The pivot point allows the mass to move along the line of motion and movement of the main gear to the side (unbalanced force).

When the wheels touch down at an angle because of misalignment or sideward drift, the pivot point is displaced from the uniform motion allowing the mass to continue moving along the direction of motion, rotating the airplane. The amount of the force on the pivot point equals the size of force, that is the speed and mass. *Kinetic energy* is related to how fast an object is moving, the faster the airplane goes, the more kinetic energy it has: Kinetic Energy = (1/2)*(mass)*(velocity)^2. Also, the direction of the force on the pivot point (to the side) is opposite to the direction of the force of motion (straight ahead).

Keep in mind that in addition to a touchdown at an angle, the swerve can also occur because of encountering bumps, chuckholes, cracks in pavement, or snow ridges; each of which can transfer the inherent instability into a swerve. Once this swerve begins it must be corrected immediately. Unchecked the airplane will swing farther and farther from the straight-line motion and the airplane, depending on its speed, may begin to tip until the outside wingtip contacts the ground. The sideward force imposes stress that may collapse the gear.

## Landing Gear Forces

During normal ground operations the surfaces of the landing gear-attach point manage the forces by distributing the force over the entire mated surfaces (*Figure 7*).

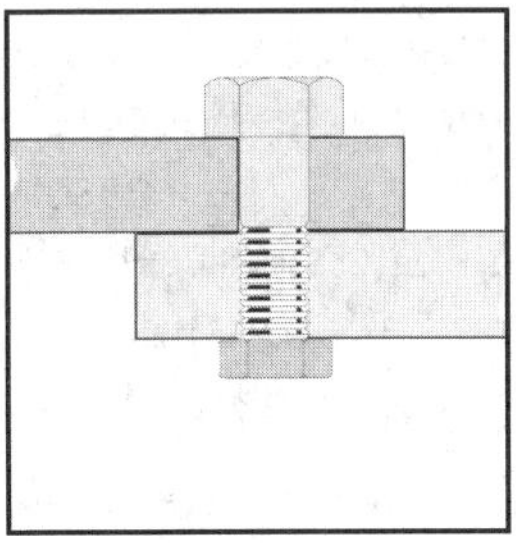

*Figure 7.* **Landing gear attach bolt.**

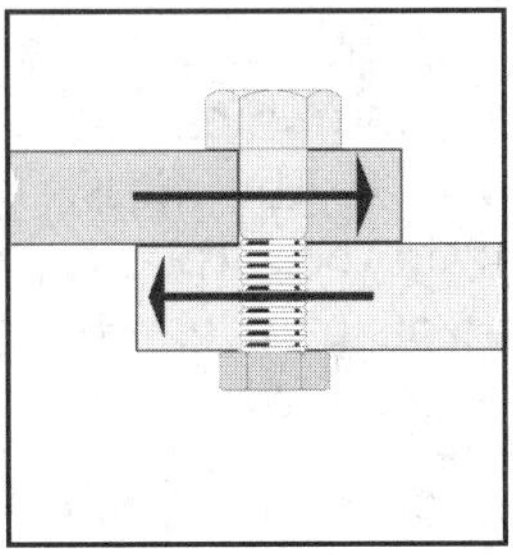

*Figure 8.* **Landing gear shear load.**

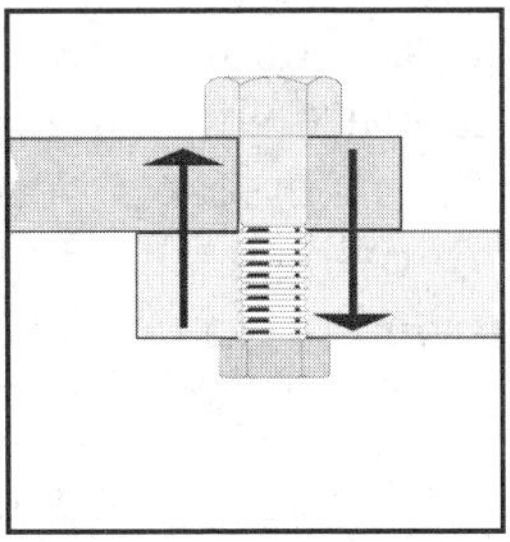

*Figure 9.* **Landing gear tension load.**

The abnormal operation of a ground loop causes shear and tension loads of the attach point. Rather than the mated surfaces absorbing the forces, the attach bolt itself is subjected to the force and may shear (cut off) the bolt or strip the attach nut from the bolt threads. Gear failure follows.

## THREE-POINT LANDING

The three-point landing is often used synomously with the term '*full-stall landing*.' It is possible to make a three-point landing without making a full-stall landing, depending on the airplane's attitude when sitting on the ground.

Often when performing the three-point landing the tail wheel will touch before the main wheels. The main wheels will drop (or be forced down). When this occurs it is desirable that the height of the main wheels above the ground be just a few inches. This occurrence, with the tail wheel touching first, is more properly called the 'full-stall landing.'

Recognizing that there is a difference, we will use the term "three-point landing" to incorporate both types of touchdown.

For crosswind operations it is important to recognize that different airplanes have different degrees of slow flight control effectiveness or authority.

If the three-point landing is used for a crosswind landing, know your airplane. The flight controls must be effective at slow speed.

Once the airplane is on the ground, contrary to the FAA admonition about not touching anything in the cockpit until after coming to a complete stop, raise the flaps as soon as possible to place more weight on the wheels. This results in better tail wheel steering and more effective braking.

### Advantages Of The Three-point Landing

- ✔ *Minimum kinetic energy (the energy of motion) at the time of touchdown which reduces the centrifugal force of any developing swerve.*
- ✔ *Touchdown in the full-stall attitude prevents bouncing or sufficient lift to become airborne once on the ground (however, gusts can change the effective relative wind to cause the airplane to become airborne).*
- ✔ *Contrary to popular belief, the tail-wheel low landing provides a shorter landing distance because more weight is on the main wheels for more effective braking.*

### Disadvantages Of The Three-point Landing

- ✔ *You are exposed to the effects of a crosswind and accompanying gusts for a longer duration at a time when the airplane is most vulnerable to gusts.*
- ✔ *Poor visibility over the nose due to the nose-high attitude.*
- ✔ *The on-the-ground angle of attack makes the taildragger have a natural tendency to bounce (actually to begin flying again) when the speed is not dissipated prior to touchdown.*

### Procedure

Make a normal 'stabilized' approach to the runway. The secret to a good approach is consistency. To make the approach become routine, fly the traffic pattern at the same altitude above the ground each time. Fly the same distance from the runway each time.

If the distance from the runway is desired to be one-quarter mile, determine a point on the strut of a high-wing airplane or on the wing of a low-wing airplane that provides the proper distance interval. Then fly this position each time.

By reducing the variables, the landing will be safer because you can concentrate on other factors, such as traffic, turbulence and the like.

Generally the pre-landing check is accomplished at the same point on each approach, that is, downwind opposite the point of touchdown.

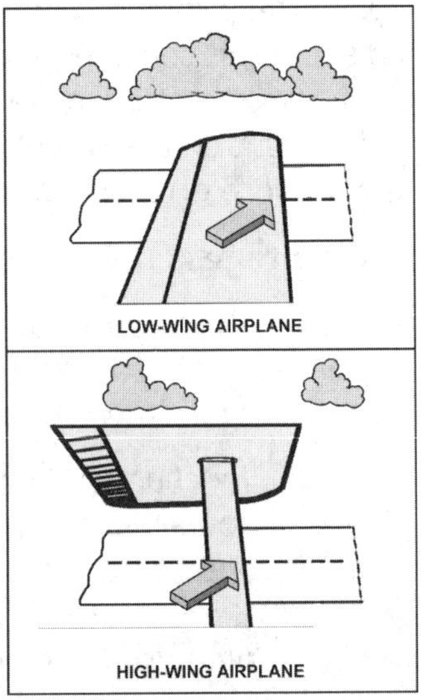


*Figure 10.* **Determine the point on the wing (low-wing) or the strut (high-wing) where the runway centerline intersects.**

**For consistency fly this same distance from the runway at all airports.**

**Mark the point with tape, by noting a distinguishing mark or create a mental image.**

Turn base when the aiming point (the point on the runway where the flare is initiated) is mid-way between the wing tip and tail.

Adjust the mixture for the landing. Many pilots who do not use a checklist were taught the pre-landing check by rote. They spout off, often without thinking of the consequences, "Fuel selector on fullest tank, mixture rich, carburetor heat on, reduce throttle, flaps as necessary, trim for the approach speed."

Moving the mixture control to the full-rich position becomes a habit pattern. When landing at airports above 3,000-feet agl (density altitude), the mixture should be adjusted to afford maximum power in the event a go-around is necessary.

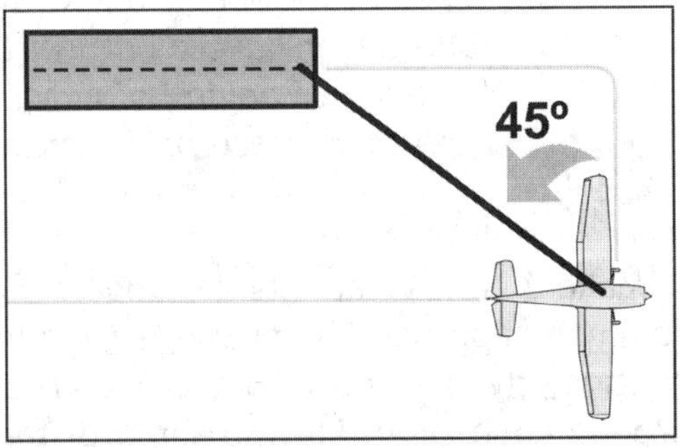


*Figure 11.* **Turn base when the aiming point is mid-way between the wing tip and the tail.**

The approach speed should be 1.3 $V_{SO}$. This provides a 30-percent margin above the stall speed in landing configuration. But, this is not the landing speed. While on short final begin a transition to 1.2 $V_{SO}$.

### Elevator Control

At touchdown the stick will be fully aft. Keep the stick in your gut during the after-landing rollout and during taxi.

## WHEEL LANDING

The wheel landing can be described as one in which the main wheels touch from any attitude and applying some forward pressure keeps the tail wheel from touching until the pilot wants it to touch.

---

**During any landing in a taildragger, if you aren't moving the rudders, you are close to a ground loop.** — *Howard Ballew*

---

There can be a tail-wheel high wheel landing or a tail-wheel low wheel landing and any range in between.

More effective braking than during the three-point landing.

The pilot can keep the tail in the air until the last minute at which time the tail is lowered to provide a rapid transition from rudder effectiveness to tail wheel steering effectiveness, lending better directional control.

### Advantages

- ✔ *Better control response to counter drift.*
- ✔ *Easier to perform a go around (especially in low-powered airplanes).*
- ✔ *Better visibility.*
- ✔ *More weight on the main wheels for more effective braking. Caution: Aggressive braking can flip the taildragger onto its back.*

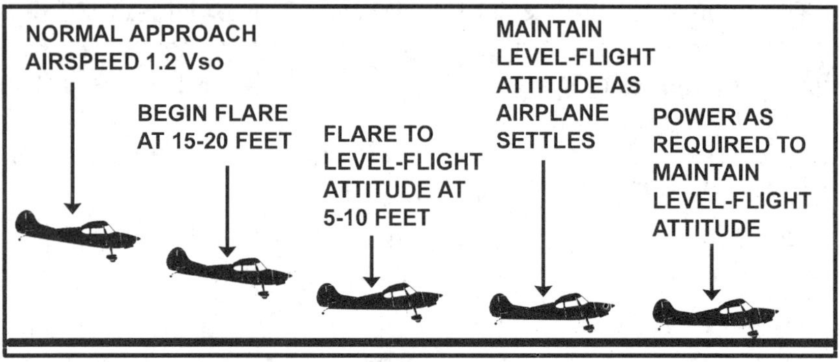


*Figure 12.* **Wheel landing. Normal approach to level-flight attitude. Slight forward stick when the wheels make contact.**

## Disadvantages

- ✔ *Greater potential for flipping upside down on a soft field.*
- ✔ *Greater potential for disaster if performed badly. Requires greater dissipation of energy over a longer period of time that can result in uncontrollable swerves.*
- ✔ *Requires recency of experience.*

## Procedure

Use the same approach speed as with the three-wheel landing. The maximum speed for maneuvering is 1.3 $V_{SO}$, then transition to the over-the-fence speed of 1.2 $V_{SO}$. For gusty winds, add one-half the gust factor.

On short-final transition to 1.2 $V_{SO}$. Continue the approach the same as the three-point landing; however, after initiating the flare to level-flight attitude carry a minimal amount of power while flying down to the runway without continuing the flare.

Small power adjustments may be required to arrest the rate of descent when just a few inches above the runway. When the wheels touch, either relax the back pressure or move the stick slightly forward to maintain contact with the runway.

The wheel-landing touchdown speed is greater than that during the three-point landing. Because the mass of the airplane lies behind the main wheels, if you do not relax the back pressure or apply slight forward pressure (depending on trim setting), the tail will continue moving downward.

If the tail movement is not stopped, it is possible and probable that when the tail makes contact there will be sufficient airspeed at the high angle of attack of ground operations to cause the airplane to become airborne again. Some call this a bounce. It's merely excess speed allowing the airplane to fly at the high angle of attack.

The brakes can be used before the tail is lowered, but use caution. The newer airplane brakes are effective enough to flip the plane on its back.

An advanced practice maneuver is to use the brakes and power to keep the tail up in level-flight attitude until the airplane approaches a complete stop.

Lowering the tail is done at the pilot's discretion. When lowering the tail don't pull back so hard and so fast that you cause damage, but make it a positive lowering. You will be transitioning from rudder effectiveness to tail-wheel steering effectiveness and any delay between the controllability transition makes the airplane susceptible to gusts with no way to compensate.

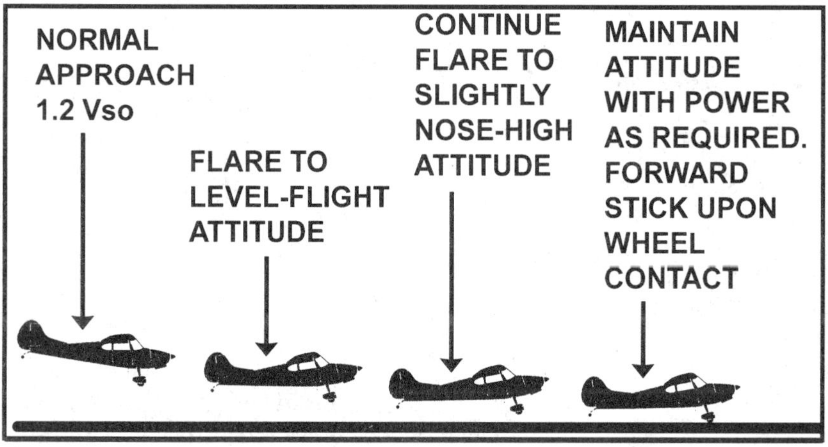


*Figure 13.* Tail-low Wheel Landing. Forward stick is required when the wheels contact the ground or the tail will continue downward.

### Elevator Control

Once you have decided to lower the tail wheel and can do so at a speed or rate that does not cause the airplane to become airborne again, pull the stick into your gut and keep it there throughout the after-landing rollout and during taxi.

## TAIL-LOW WHEEL LANDING

The tail-low wheel landing comes in an infinite variety since any attitude with the tail below the level-flight attitude can be defined as the tail-low wheel landing.

### Elevator Control

Lowering the tail from the tail-low wheel landing is at the pilot's discretion. Be sure the speed is slow enough so the airplane does not become airborne again. Pull the stick into your gut and keep it there throughout the after-landing rollout and during taxi.

---

**After about 30 minutes I puked all over my airplane. I said to my self, "Man, you made a big mistake."** — *Charles 'Chuck' Yeager, regards his first flight*

---

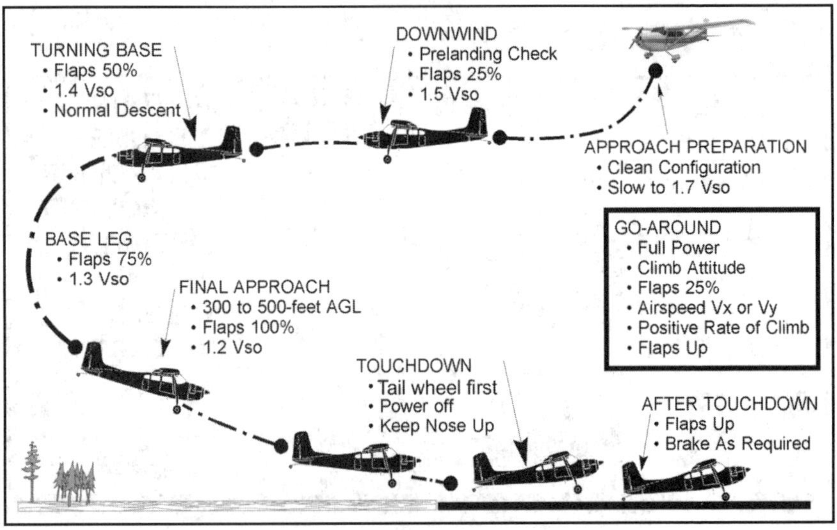


*Figure 14.* **Soft-field landing.**

### Practice Maneuver

A valuable practice maneuver for the three-point landing, wheel landing or tail-low wheel landing is to make a normal approach for landing. Once the flare has been initiated, or the level-flight attitude attained just inches above the ground, add power and continue to fly the length of the field without touching down.

## SOFT-FIELD LANDING

The soft-field landing is a technique that is used on either soft fields or rough fields. It is the preferred method for landing on snow-covered or slush-covered runways.

Do not make a long, flat approach at slow airspeed. Use the spot method to arrive at the aiming point from a normal approach.

The only difference between a normal landing and the soft-field landing is the final portion of the approach and the touchdown.

Perform a normal approach for the landing. On downwind, opposite the landing point, perform the pre-landing checklist. As the speed slows below the maximum flaps extended speed, apply 25-percent flaps.

Most POHs or owner's manuals do not provide a caution about adding flaps while in a bank; however, it is the sign of a professional pilot to apply flaps while in wings-level flight. There are two reasons for this. First, when the airplane is banked, g-force is encountered that places additional stress on the flaps. Second, if the flap mechanism malfunctions and a split-flap condition occurs, it is much easier to control the airplane in wings-level flight.

---

**Experience is a hard teacher. First comes the test, then comes the lesson.** – *Aviation Cliché*

---

Prior to the turn to the base leg, apply flaps for a total of 50 percent. After completing the turn onto the base leg, apply flaps for a total of 75 percent. The airspeed should be 1.3 $V_{SO}$.

When on final approach, full flaps are extended with the airspeed 1.3 $V_{SO}$. On short final the airspeed is slowed (by pitch adjustment) from 1.3 to 1.2 $V_{SO}$. During the flare the power is increased in an attempt to keep the airplane from touching down.

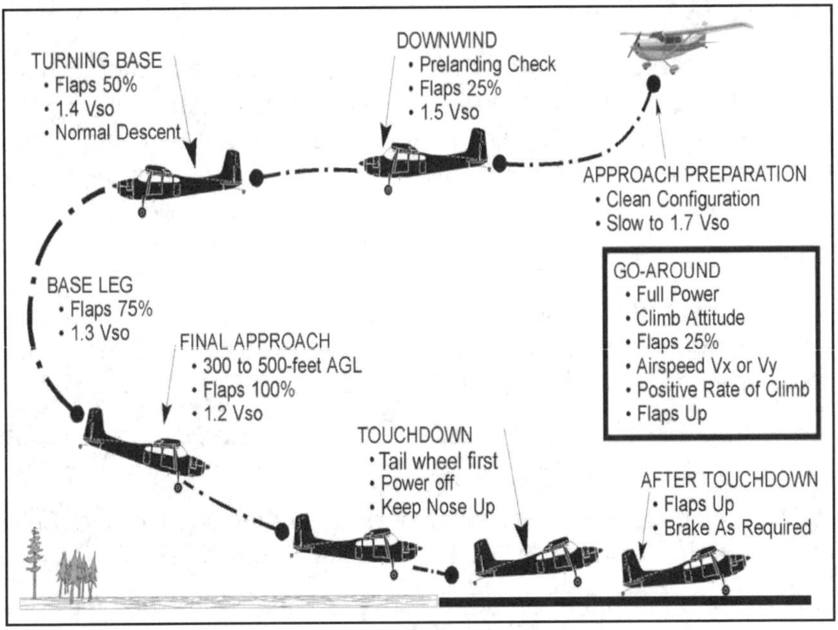


*Figure 15.* **Short-field landing.**

In an airplane with a fixed-pitch propeller, the power change will be in the range of an increase from 1,500 rpm to 1,800 rpm. This allows the touchdown to be made at a speed much slower than usual.

Touchdown is made in the three-point attitude. When the main gear touches, reduce power to idle and pull the control wheel full aft. Retract the flaps to place more weight on the main wheels and tail wheel. This will also help in preventing damage to the flaps if mud and rocks are slung by the wheels.

If the surface is soft, as opposed to rough, do not come to a complete stop until reaching the parking area. Otherwise it will take excess power to start the airplane moving again.

## SHORT-FIELD LANDING

Proficiency in short-field operations is required before trying to operate at the short field. The FAA's definition of a short field includes a 50-foot obstacle at the end of the runway. This is the reason for the short field; the tree makes a portion of the runway unusable. Many short fields do not have the obstacle approach, but they can be short.

The most important aspect of the short-field landing is airspeed control. The proper approach speed must provide for safe maneuvering margins, gusts and wind shear, and flare capability.

### Safe Maneuvering Margins

The initial approach on the downwind leg is made at 1.3 VSO. This provides a 30-percent margin above stall speed for maneuvering. The bank in the traffic pattern is normally limited to 30 degrees (a 30-degree bank - 1.1547 load factor - results in a 7.45-percent increase in stall speed), so the 30-percent margin is adequate. (A 30-percent increase in the stall speed occurs at around 54 degrees of bank.)

The stall speed increases as the square root of the wing load factor. In a 60-degree bank there is a 2-g load factor. The square root of 2 is 1.41, or a 41-percent increase in stall speed.

### Gusts and Wind Shear

To compensate for gusts and wind shear, add one half the gust factor to the normal approach speed.

### Flare Capability

Each airplane has a limited amount of elevator control available for the pilot to flare. For example, the elevator of the Cessna 170B has 26 degrees of nose-up pitch movement. As the airspeed slows, the amount of elevator control must be increased to maintain a constant attitude.

If the airplane is allowed to approach the runway at an airspeed that is too slow, there may be insufficient elevator control to effect the flare. The airplane will continue its rate of descent until it strikes the ground. If this happens, add power so the prop blast increases the elevator efficiency and the additional power will decrease the sink rate to cushion the landing.

### Airspeed is All Important

REMEMBER

The landing distance increases by the square of the ratio of the touchdown speed to the normal touchdown speed.

Assume the normal approach speed is 50 knots. For some reason we get sloppy and approach at 55 knots instead. The 5 knot increase in approach speed doesn't seem like much, but the ratio of

55/50 is 1.1, or a 10 percent increase in the approach speed. The square of the ratio is 1.21. This relates to a 21 percent increase in landing distance.

Some pilots might argue, "My airplane requires 1,000 feet to land, so what do I care if it takes 1,210 feet? That's not such-a-much."

With this attitude, complacency rather than common sense takes over. So what happens the next time when the approach speed is 70 instead of 50? 70/50 is 1.4 or a 40 percent increase in approach speed. 1.4 squared is 1.96, or a 96 percent increase in landing distance.

It is not ballpark airspeed control that is important for the short-field landing. It is *exact* airspeed control.

We have proved a small increase in approach speed equals a large increase in landing distance. It should be kept in mind the airplane will not land as short at high-elevation air strips as it does at sea level. The same IAS is used at sea level and at high elevations, but the TAS increases about two percent per thousand feet above sea level.

---

**Don't drop the aircraft in order to fly the microphone.** – *Aviation Cliché*

---

## Short-field Landing Technique

The short-field landing can be defined as one where the airplane touches down on a precise spot at minimum speed, in a full-flap, nose-high attitude with little or no float, from a controlled descent.

There are two facets of the short-field landing that will make or break the landing. The first we have discussed, airspeed control. The second is the approach. If the spot method is incorporated in the approach, short-field operations become a "piece of cake."

- *Begin slowing the aircraft to 1.3* $V_{SO}$ *upon entry to the pattern.*
- *Downwind opposite the point of touchdown, perform the pre-landing cockpit checklist.*

- *Add 25 percent of the flaps opposite the touch-down point (airspeed below the maximum flaps extended speed).*
- *Prior to or after turning base, apply 50 percent flaps.*
- *After turning final apply full flaps.*
- *On short final, approximately 200 feet agl, transition to 1.2* $V_{SO}$. *Begin the flare at five- or 10-feet agl.*
- *After touchdown, move the control wheel full aft, retract the flaps and brake as required.*

If an obstacle is present that requires a steeper approach angle, use the spot method for the 4.5-degree approach angle. It is poor pilot technique and the source of a potential accident to "chop" the power after passing the obstruction to drop down to the runway. A constant descent angle is maintained from final approach to the flare (aiming point).

Suppose the approach appears to have gone properly, but during the flare the airplane floats and floats. You have determined ahead of time that there is no possibility of a go-around at this strip. What do you do?

There are two possible choices, but do not push the nose forward to land. This will most likely cause a bounce. When the airplane is out of control, there may be no chance of stopping in the space available.

When close to the ground, get rid of the flaps. The coefficient of lift will be reduced so the airplane can land. Do this close to the ground, not five feet in the air. If maximum braking is required, apply pressure up to the point where the wheels are just about to skid the tires, but do not lock the brakes. Besides wearing off one side of the tires, it will take longer to stop. Skidding tires produce less resistance than when they are rolling with hard braking action.

---

**Flying is not Nintendo. You don't push a button and start over.** — *Aviation Cliché*

---

The other choice is a forward slip. This should be practiced with an instructor. It will not take long to become proficient in a forward slip to a landing. For practice, approach with 10-15 knots excess approach speed. Begin the forward slip about 50-feet agl and carry it into the flare. When the airspeed dissipates sufficiently to allow a landing without further floating, release both aileron and elevator controls at the same time. The longitudinal axis of the airplane will parallel the runway and a normal three-point landing may be made.

If the airplane is floating in the flare, begin a hard forward slip. Usually the pilot will press full right rudder and compensating left aileron. As the speed slows, relax both controls at the same time. The reason for using the right rudder is to move the nose right and to provide the pilot with better visibility in those airplanes with side-by-side seating.

### Safe Landing Principles

- *During approach, line up with the runway center line and stay lined up during the approach, touchdown and rollout. A crab is used until about 100-200 feet agl, then transition to a side slip.*
- *The slip method of alignment is preferred for a crosswind landing. If the airplane cannot maintain runway centerline alignment below 20- to 30-feet agl, the wind is too strong to safely land.*
- *Always determine a land/no-land point on the runway. If you determine you are overshooting, go around.*
- *Clear the area for traffic constantly. The final approach area has a high potential for mid-air collisions.*
- *Use the spot method to avoid dragging in low on final approach or diving on the airport.*
- *Use common sense and sound judgment.*

## LANDING IRREGULARITIES

It doesn't matter how good you are, or how good you think you are, executing a perfect flare and landing is not possible each time you approach your destination airport. The taildragger pilot must be aware of how to take care of his mistakes.

## GUSTY WIND CONDITIONS

When the steady-state wind is plagued by gusts, add one half the gust factor to the computed approach speed to compensate for airspeed changes caused by wind shear.

For example, if the wind is 10 knots gusting to 20 knots, the gust factor is the difference between the two values or 10 knots. Add half this value, five knots, to the calculated approach speed.

## AIRSPEED INDICATOR ERRORS

For newer airplanes, those manufactured since the mid-1970s, the airspeed indicator is calibrated in indicated airspeed. The standard color-code markings denote operating limitations in indicated airspeed.

For airplanes manufactured before the mid-1970s, operating limitations, such as the color-coded face of the airspeed indicator, are calibrated airspeed.

Calibrated airspeed is the indicated airspeed corrected for installation error and instrument error. The installation error is greatest at low airspeeds and the instrument error may be as great as five or 10 knots. At cruise airspeeds the installation error (or position error) is negated and the indicated and calibrated airspeed are approximately the same.

It is important, in airplanes that use the calibrated airspeed method, to refer to the airspeed calibration chart in the Owner's Manual to correct for airspeed errors before using any particular airspeed that defines an operating limitation, such as flap operating speeds, maneuvering speed and the like.

### Airspeed Calibration Chart

For those airplanes with operating limitations published as calibrated airspeed, always use the calibrated airspeed to calculate the proper speed at the operating weight being used. Convert this calibrated airspeed to indicated airspeed and use this value for adhering to the operating limitations.

### Reduced Landing Weight

If the Pilot's Operating Handbook or Owner's Manual does not list the airspeed for operating at reduced weights, a rule-of-thumb may be used. Reduce the approach airspeed for the maximum landing weight listed by one-half of the percentage of the weight decrease.

To determine the percentage of weight decrease, subtract the actual weight of the airplane from the maximum allowable landing weight. Divide this value by the maximum allowable landing weight.

For example, assume the maximum landing weight of the airplane is 2,300 pounds. If the airplane is being operated at a reduced landing weight of 2,000 pounds. We find the percentage of weight decrease, by subtracting 2,300-2,000 = 300. Then, 300/2,300 = 0.13, or 13 percent below maximum authorized landing weight.

The airplane's landing weight is 13 percent below maximum. The approach calibrated airspeed should be reduced by one-half of this value, or 6.5 percent.

Using a maximum landing weight approach speed of 65 KIAS, multiply by 6.5 percent (0.065) and subtract this value from 65 KIAS. This gives us a reduced-weight approach speed (65 x 0.065 = 4.225. 65 - 4 = 61 KIAS).

## BALLOONING

When the apparent upward movement of the ground is misjudged, and it appears to be rising faster than it actually is, there is a tendency to increase the pitch attitude too rapidly. This will not only cause the aircraft to stop descending, but to start climbing. This may be caused by concentrating the vision too close to the aircraft, resulting in a "speed blur" that makes it impossible to judge the aircraft's height above the ground. Ballooning is dangerous because the height above the ground is increasing and the airplane may be approaching a stall.

## Corrective Action - Ballooning

When ballooning is slight, maintain direction control, hold a constant landing attitude, and add a small amount of power to allow the airplane to settle onto the runway.

After determining the severity of this error, the use of the throttle is increased with increasing altitude above the surface to cushion the landing; the throttle must be reduced immediately after the touchdown.

When ballooning is severe, make a go-around immediately. Do not attempt to salvage the landing. Saving a landing from a bad balloon condition requires an extremely high degree of judgment and proficiency, with which, the balloon probably would not have occurred. If in doubt, do not take a chance, go around.

We just discussed "attitude" in the previous section on Simplified Landing. Two attitudes were presented as important to the landing; level attitude and climb attitude. Either a slight or severe balloon requires the pilot to transition to the level attitude immediately. If the landing is to be continued, transition slowly to the landing attitude (climb attitude). If a go-around is executed, transition to the climb attitude as airspeed increase allows.

## BOUNCING

Recovery wise, a bounce is very similar to ballooning; the only big difference is the initial causal factor. If the aircraft strikes the runway severely, it will bounce into the air, just as a ball will bounce. The height it reaches depends on the force that it strikes the runway and may be aggravated if the pilot is startled and pulls back on the control wheel. This is the speed at the point of the touchdown, or, for a stall, the height the airplane falls.

---

**It doesn't do any good to stand on the airplane's brakes when you're already on your back!** – *Rex Thorp*

---

Since a bounce occurs if the aircraft makes contact with the ground before the landing attitude is attained, it is almost invariably accompanied by excessive back pressure. This is the result of realizing too late the aircraft is not in the landing attitude and attempting to establish the attitude just as touchdown occurs.

Or, a normal protective reaction is to pull back on the control wheel when the airplane makes ground contact, causing the bounce.

Another cause of the bounce is touching down prematurely with excess speed. Because of the high angle of attack when the airplane is on the ground, excess airspeed allows the airplane to become airborne again. This is often referred to as a bounce, but is actually a touchdown with excessive airspeed.

### Corrective Action - Bouncing

The corrective action for a bounce is the same as for ballooning and depends on its severity. When the bounce is very slight and there is no abrupt change in pitch attitude, a follow-up landing may be executed. Maintain directional control, apply power to cushion the landing, and smoothly adjust the pitch to the landing attitude before the subsequent touchdown. Remember, keep the control wheel (stick) back. Too many pilots try to continually adjust the flare and set up a porpoise effect that continues throughout the series of landings that follow. This is called PIO (pilot induced oscillation).

A severe bounce, one where the aircraft rises too rapidly, requires an immediate go-around. Do not attempt to salvage the landing. Simultaneously apply power and lower the nose to level attitude or slightly above level attitude. Follow through with the go-around procedure although another bounce may occur. As the airspeed increases, gradually increase the pitch attitude to the climb attitude.

- ✔ *Light Bounce - Less than one-wing span above the ground. Pull the stick, at the appropriate rate, into the gut and keep it there.*
- ✔ *Moderate Bounce - One-wing span above the ground. Resume the landing attitude, cushion the sink with power and pull the stick into the gut and keep it there.*

✔ *Severe Bounce - More than one-wing span above the ground. Consider making a go-around. Pitch to level-flight attitude, add power, once the airplane is controllable reduce power and make a three-point landing.*

## PORPOISE

The porpoise is the least expected of all aircraft landing incidents—that can lead to an accident. Some pilots do not know what the porpoise is, how to prevent it, or the proper corrective action. Even experienced pilots may fly for years without experiencing the porpoise, but, when they do, they will not soon forget it.

Perhaps the most common producer of the porpoise is the bounced landing. Because of distractions, reduced visibility or complacency, the airplane is allowed to bounce. This may startle the pilot so he overreacts and pulls on the elevator control. This aggravates the situation. Seeing the airplane rising rapidly away from the ground, he shoves the elevator forward in an attempt to stop the climb. The back elevator pressure needed before the next impact always comes a moment late and the airplane impacts the runway, and then climbs again. This process continues with each ensuing bounce becoming worse.If the pilot induced oscillation is allowed to continue it may lead to the collapse of the main gear or a prop strike. It won't stop until back pressure on the elevator is held constant stopping the porpoise.

---

**I am not afraid of crashing, my secret is ... just before we hit the ground, I jump as high as I can.** — *Bill Cosby*

---

Another causal factor is too much airspeed during the landing. With diminishing runway available for stopping, sometimes the airplane is helped with its landing with a little forward pressure on the control wheel. This may cause the porpoise.

### Corrective Action - Porpoise

Once the porpoise begins, few experienced pilots can do a pumping action of the elevators with the precision required to stop the porpoise. With more than 20,000 flight hours in small planes, the author is still unable to master this technique with any degree of regularity.

The only solution is to change the pitch to level-flight attitude, cushion the sink with power and as the airplane approaches the ground, assume the landing attitude and bring the stick back into your gut.

Checking out pilots—including a few high-time airline pilots who have not been in a small airplane for years—can scare an instructor during landings. At some altitude, say 50 feet above the ground, they may begin the flare. Although it is too high, it is OK, until they continue the flare at a rate too fast. This leads to running out of elevator control before reaching the ground ... making quite an arrival if the instructor doesn't get on the controls rapidly. This is the true full-stall landing, but not the kind desirable for safety.

---

**When once you have tasted flight, you will forever walk the earth with your eyes turned skyward, for there you have been, and there you will always long to return.** – *Leonardo da Vinci*

---

The "perfect" landing consists of a level off to level-flight attitude somewhere around 5 to 15 feet above the ground. Even if it is higher, there is no particular problem. If the flare to landing attitude is performed too quickly from this higher altitude, it is necessary to stabilize the airspeed decay with the addition of power and with a transition back to the level flight attitude. The flare is not continued until the airplane decelerates and begins to settle toward the runway. The flare (roundout) is continued to the landing attitude before impact with the runway.

Corrective action for a porpoise that results in the airplane bouncing or being flown more than 15 feet above the runway, is to add full power, maintain directional control, and transition to level attitude. As the airplane accelerates, make a slow change to the climb attitude and execute a go-around.

If the airplane bounces fewer than 15 feet, do not shove the control wheel forward in an attempt to land the airplane. If the control wheel is adjusted, do not lower the nose beyond the level flight attitude. Should you make the decision to land, normally, by increasing back pressure on the elevator control, at a rate that does not produce a balloon, the airplane will go through a series of bounces that get smaller and smaller. With increasing back pressure, one or more bounces may occur, but aircraft damage is not as likely as with a continuing porpoise condition.

Do not try to recover from the porpoise with forward elevator movement. Always make a go-around or recover to the level-flight attitude followed with back pressure (and sometimes power to cushion the landing), depending on the severity of the bounce.

**If God had really intended men to fly, He'd make it easier to get to the airport.** — *George Winters*

## GROUND LOOP

A ground loop is an uncontrolled turn during ground operations that may occur while taxiing or taking off, but especially during the after-landing roll. The conventional gear airplane is susceptible to the ground loop. It is not always caused by drift or a weather vane effect although these things may cause the first swerve. Careless use of the rudder, a flat tire, brake failure, uneven ground, or a soft spot that retards one wheel may also cause a swerve. In any case, the initial swerve only makes the aircraft ready for the ground loop.

Due to the structure of an aircraft with conventional landing gear, the forces that cause a ground loop increase as the swerve increases. The first swerve develops centrifugal force and this, acting through the center of gravity located behind the main gear, swerves the aircraft even more. Centrifugal force increases as the square of the speed where it begins.

If allowed to develop, the centrifugal force produced will become great enough to cause the airplane to swing around. At that point you are along for the ride. The runway surface and the speed at which the ground loop occurs will determine the amount of damage to the airplane.

At slow speeds (10 knots or less) the airplane may, in the blink of an eye, swap ends in a pirouette that would be the envy of any ballet performer.

A faster speed (20 knots) may tip the aircraft during the rapid whirling until one wing tip strikes the ground.

Faster yet (30 knots or more) and the landing gear will experience shear loads, then tension loads that will break the attach bolts and the gear folds under the airplane. When this occurs it will obviously damage the wing too.

### Corrective Action - Ground Loop

The use of power to counteract the centrifugal force developed during the swerve, along with coordinated use of rudder, aileron and brake is the only remedy for stopping the incipient ground loop.

> *Once you recognize the beginning of a ground loop, add full power and raise the tail wheel. The tail must be up to transfer the mass of weight from behind the main gear to the main gear.*

If the swerve is to the right, some more experienced pilots will use full right aileron control (left aileron deflected down, right aileron deflected up), in an attempt to lift the left wing. Also, the increased lift on the left wing will increase drag.

Depending on ground speed, the aileron opposite the swerve may be more effective. Here, full left aileron will produce a down deflection on the right wing. This will produce lift, helping to turn the airplane to the left, in a direction opposite the swerve. If the aircraft has swerved so far that making wing contact with the ground is likely, this procedure will aggravate the situation and may cause the wing to touch. It must be accomplished at the time of the initial veer.

## SIMPLIFIED LANDING

Reminiscing about days past often provides one insight when analyzing the various flight maneuvers and procedures that do not turn out quite as perfect as one's expectations. Although my logbook shows more than 10,000 landings, not all have been "greasers."

We all make excuses for a botched, or at least a less-than-perfect landing. My favorite justification is to tell myself that I constantly fly different makes and models and the visual perspectives change for each airplane.

Excuses are just that. When I find myself making excuses, it is time to analyze what is happening. Reverting to my primary instructor days helps toward this end. Contemplating how landings are taught usually provides the insight to solve what has gone wrong.

*Figure 16.* **Level flight attitude. If the natural horizon is not visible, use the base of the mountains about 6-8 miles away.**

It is said the approach is all-important to a good landing. What is a good approach? An approach is good when the glide path control leads to the aiming point on the runway while maintaining exact airspeed control. *See: Spot Method for Landing*

While few pilots will disagree that a good approach is necessary for a good landing, there is more to it than that. It does not matter what kind of approach is made if there is no flare leading to an acceptable touchdown.

*Figure 17.* **Climb attitude. Note where the horizon (base of the mountains) intersects the cowling.**

Often a bad approach can be salvaged—if it does not involve a compromise of safety—by making the appropriate adjustment for being too high or too low. The flare and touchdown are the secret.

Regardless of the approach for landing, whether it is a no-wind landing, a crosswind landing using a crab or a side-slip technique, it is ALL important that the airplane touchdown straight with the longitudinal axis perfectly aligned with the movement of the airplane. Drift, the sideward movement across the runway, or crab, the angular correction of flying into the wind to prevent drift, cannot occur when the wheels touch the surface.

*Figure 18.* **Glide attitude. The "+" represents the windscreen mark for level flight attitude. Note the position of the natural horizon (base of the mountains.**

One way to make consistently good landings, especially when changing from airplane to airplane, involves basic attitude flying; that is, the relation of the nose of the airplane to the horizon. Flying in mountainous terrain requires one to use the base of the mountains some six to eight miles away as the natural horizon.

To develop the sight picture of the required attitudes for making a perfect landing, climb to a safe altitude. First, determine the attitude for level flight. Look at the horizon and notice where it intersects the windshield. This will probably be about one to four inches up from the base of the windshield.

Next, learn the attitude for climb at the best rate of climb airspeed. The horizon will intersect the side of the cowling below the nose. Memorize the position of the nose with respect to the horizon for these two attitudes. These

*Figure 19.* **Approaching about 30 feet above the runway.**

are the level attitude and the climb attitude.

Next, cover the airspeed indicator, and make the transition from level attitude to climb attitude. Check the airspeed indicator. If the airspeed is not within one knot of the best rate-of-climb speed, practice some more. Change back to level-flight attitude. Check the instruments to see the indications show level flight.

*Figure 20.* **The flare to level flight attitude is begun about 10-15 feet above the ground.**

Practice these transitions—from level-flight attitude to climb attitude and back—until the airspeed can be nailed within one knot. It will take much less time than one imagines, five to 10 minutes at most.

*Figure 21.* **The flare is continued to the level flight attitude.**

Put it together by going through a pre-landing check and establishing the normal approach airspeed. Trim the airplane to maintain the approach airspeed. Learn this approach (or glide) attitude. Practice the flare to level-flight attitude, pause, then continue the flare to the climb attitude. This practice should be accomplished with and without flaps.

*Figure 22.* **Continue the flare to the climb attitude.**

Move to the traffic pattern. After making the perfect approach for the landing, transition to the level-flight attitude at five to 20 feet above the runway. When sinking is detected, make a slow transition to the climb attitude. The transition to the climb attitude must be made at a rate that will not cause a balloon. The climb attitude must be reached before the actual touchdown, but not while the airplane is more than one foot above the runway.

Students, having a hard time developing the perspective on height above the runway, will find this technique helps establish the viewpoint necessary for landing.

Experienced pilots will find this technique valuable in eliminating the "thumpers" that inevitably sneak up on us all.

## SUMMARY

The pilot is well-advised to incorporate the spot-method, using the stabilized approach, for all landings. Learning and practicing this technique will pay dividends throughout the years.

Proficient pilots usually employ the wheel landing for crosswind operations with flaps up or limited to 25 percent extention. Recency of experience counts a great deal in using the wheel landing and pilots out-of-practice normally use the three-point landing.

As an advanced-pilot technique the tail-low wheel landing will result in a shorter landing than will the three-point landing. This is due to the fact that more weight can be placed on the main wheels sooner to make them more effective.

The pilot must keep the airplane straight during landing. Any drift relative to the overall motion of the airplane will cause the airplane to swerve. Unlike a figure skater who can extend her arms to reduce the moment of inertia, the taildragger, once moving beyond a certain angle, becomes impossible to stop unless there is sufficient speed or power to raise the tail and get the weight back on the main wheels instead of behind them.

# CHAPTER 9

# Crosswind Operations

## INTRODUCTION

Approximately one-half of all aircraft involved in ground incidents or accidents during crosswind operations are equipped with a tricycle landing gear. Pilots assume that tricycle-gear airplanes are easier to operate in a crosswind, and the implication is, that taildragger pilots don't attempt crosswind landings in wind conditions as strong. We know this isn't true, so it must be that tricycle-gear pilots become complacent or have not received adequate instruction in the art of crosswind landing. Regardless of the cause, the statistics point out the need for crosswind-operation proficiency.

It is preferable to land directly into the wind, to eliminate the side slip or kick-out method that is used to compensate for the wind's sideward drift of the airplane. In the early days of aviation, when an aerodrome consisted of a large square area suitable for operations in any direction, it was possible to direct the airplane's nose into the wind. If a wind shift yaws the airplane to a new heading, the airplane could continue its operation. Today, we know, there are limited runways. Given the fickle nature of the wind, crosswind proficiency is required.

Improper crosswind technique results in a number of accidents each year, some involving fatalities. The following review is helpful to eliminate complacency, while providing a review of the principles and techniques for crosswind operations.

The crosswind-operations section will cover the following items:

- *Crosswind taxiing*
- *Crosswind takeoff*
- *Crosswind landings*
- *Crosswind after-landing roll*
- *Crosswind landing errors*
- *Maximum crosswind component*
- *Determination of the crosswind component.*

## CROSSWIND TAXIING

### Control Positioning During Taxi

Improper ground handling of an airplane under the influence of a surface wind can result in a tendency to weather cock, ground loop, nose over or to tip the airplane onto its back.

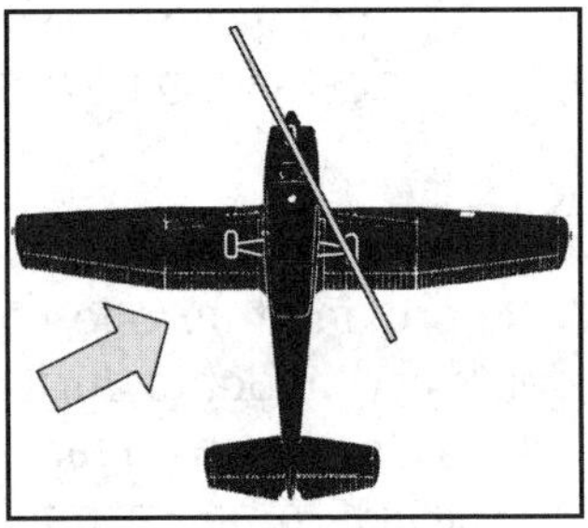

*Figure 1.* **Tricycle-gear pivot point.**

When taxiing in a crosswind, the wind flows under the upwind wing and tries to lift the airplane. Depending on the leverage advantage of the wind in relation to the physical configuration of the landing gear, the wind may succeed in flipping the airplane.

*Figure 2.* **Taildragger pivot point.**

Tricycle-gear airplanes have a physical configuration of the gear—with the nose wheel in front of the main gear—that provides the wind a leverage advantage when the wind is quartering from the rear. That is, a fulcrum or pivot point.

Conventional-gear airplanes have the steering tail wheel in the rear. This provides a leverage advantage for the wind when the wind is quartering from the front.

Regardless of the type of airplane operated—tricycle gear or conventional gear—the controls should be positioned to negate the effect of the wind. The ailerons are positioned to increase lift and drag on the downwind wing when taxiing into a quartering head wind, or to block the wind from flowing under the upwind wing when taxiing with a quartering tailwind.

Some airports, those experiencing heavy annual rainfall, have incorporated a berm (hump or mound) down the center of the taxiway. This incline to either side of centerline, that facilitates water removal, can cause problems when taxiing in a crosswind. If the airplane is taxied along the centerline or the downwind side of the taxiway, it is susceptible to more wind under the upwind wing. It is better that the airplane is taxied along the upwind side of the taxiway with the upwind wing tilted downward. This reduces the effect of the crosswind.

Exercise caution when the crosswind is 30 percent of $V_{SO}$. It is recommended, when the crosswind exceeds 50 percent of $V_{SO}$, that taxi and flight operations be suspended until conditions are more favorable.

## THUMBS-UP METHOD — FOR POSITIONING THE FLIGHT CONTROLS

It is not necessary to memorize how to position the flight controls for the different wind variations. And, it is not necessary to reason it out while taxiing and turning. The attempt can lead the inexperienced pilot to moments of anxiety. A faster, and arguably better, technique is to use the thumbs-up method.

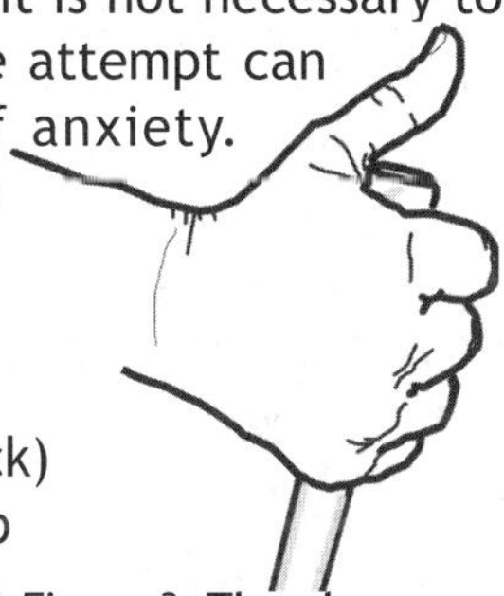

*Figure 3.* **Thumbs up.**

### Aileron Positioning

While holding the control wheel (or stick) in the neutral position, stick your thumb up vertically. Regardless of the wind direction, from front or rear, or from either side,

*Figure 4.* **Thumbs-up method for control positioning while taxiing in wind.**

merely parallel the wind with the thumb forming a straight line from the tip of the thumb to the heel of the palm.

The left-quartering tail wind and the right-quartering headwind requires the same control positioning. With a left-quartering tailwind, the wind will flow beneath the left wing trying to tip the airplane. A right-quartering head wind flows under the right wing providing lift. Because the left main wheel and tail wheel form a pivot point perpendicular to the wind, it is easier for the quartering headwind to tip the conventional-gear airplane than with the quartering tail wind.

With a right-quartering tail wind, the right main wheel and tail wheel form the pivot point, but the left main wheel opposes any tipping action.

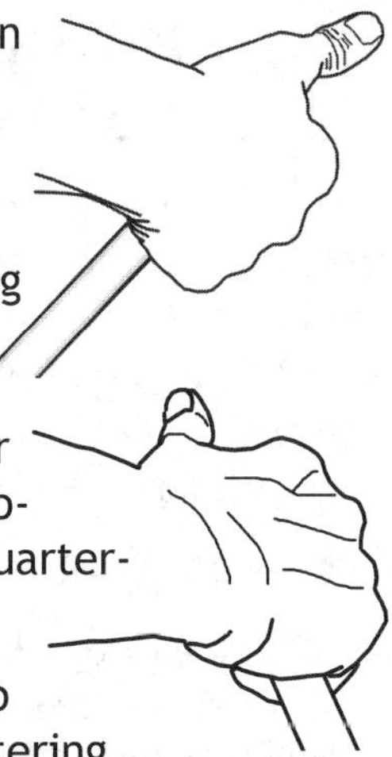

Figure 5.

In either case (left-quartering tail wind or right-quartering head wind), paralleling the wind with the thumb results in the stick being moved to the right. This moves the left aileron downward (to block the airflow of the left quartering tail wind from getting under and lifting the wing) and the right aileron upward (to reduce the lift caused by the right quartering head wind).

Similarly, paralleling the wind with the thumb in a left-quartering head wind or a right-quartering

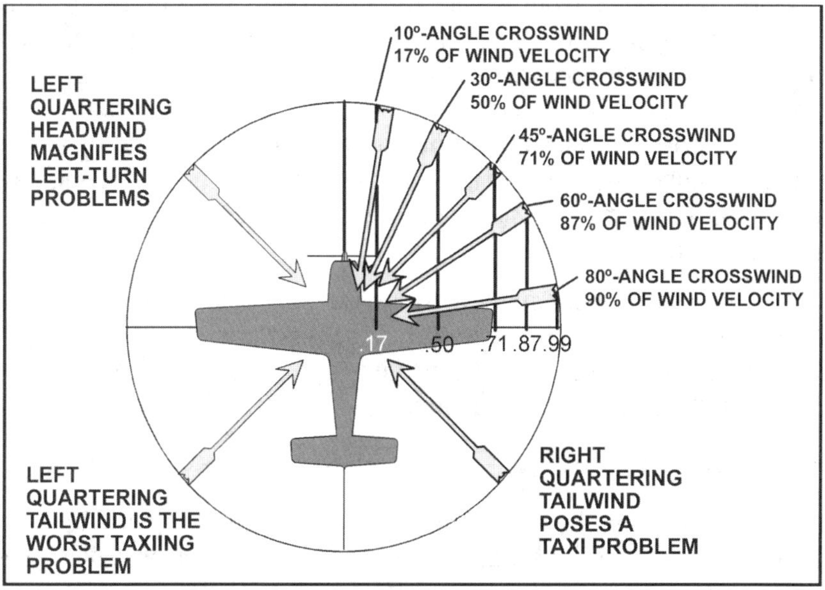


*Figure 6.* Crosswind effect and the crosswind velocity based on the angle of the wind.

tail wind causes the control wheel to move to the left, automatically placing the ailerons in the proper position.

The advantage of the thumbs-up method is that it does not require a great deal of thinking to position the controls by analyzing the movement of the stick and the resulting movement of the ailerons. Try the thumbs-up method while doing a clearing turn when the wind is blowing. This is the time it eases the cockpit workload. Paralleling the wind with your thumb automatically compensates for the wind while eliminating the need to stop and think.

## Elevator Positioning

Whenever the wind is blowing from behind the taildragger, we have been taught to move the control wheel full forward, causing the elevator control surface to move downward. This is supposed to help eliminate the possibility of wind getting beneath the tail and lifting the airplane. Unless the wind is more than about 10 knots, you will probably want to keep the stick in your gut since forward pressure reduces the effectiveness of the tail wheel steering by removing some of the weight from the tail.

When operating from a soft field or rough field, the elevator control is positioned to neutral to prevent rocks and debris from being blown back and damaging the elevator surface. When taxiing on soft or rough fields, maintain taxi speed with sufficient power so the airplane does not come to a stop. If the airplane stops, a higher power setting will be required to get it moving again, adding to the possibility of picking up rocks that may damage the propeller or tail.

In a no-wind condition or with a wind approaching the airplane from the front the stick should be "sucked back into your gut." This keeps the airplane on the ground, and generally provides better steering through the tail wheel.

## DOWNWIND TAXIING

Because of the physical configuration of a taildragger's landing gear, there is less leverage advantage for quartering tail winds to tip the plane over when compared to a tricycle-gear airplane; however, strong quartering tail winds require extra caution. There is more of a tendency to weather vane when the airplane is exposed to a tail wind. This is due to the reduced effectiveness of the flight controls.

It is easy to acquire excess taxi speed given the extra boost of the wind. If a turn is made onto the runway, without stopping or slowing, it can result in tipping over. Fuel sloshes to the outside of the turn creating centrifugal force to add to the tipping tendency. Eliminate this risk of tipping by avoiding sudden bursts of the throttle, sharp braking and excess speed.

*Figure 7.* **Highway sign.**

The tailwind also makes the airplane more susceptible to nose over. Taxiing downwind requires a speed less than normal taxi speed. A rule-of-thumb is to maintain positive control. That is, maintain the ability to stop or turn where and when necessary. While taxiing, try to make all turns into the wind, slowing before initiating the turn to take advantage of the weather vane tendency. A turn around against the wind may require full rudder, brake application and power.

Airplanes equipped with narrow landing gear are more susceptible to tipping or dragging a wing during a high-speed turn.

## Caution

Taxiing into a head wind of 15 knots or greater and then turning around (or when taxiing with a tail wind of 15 knots or greater) requires the use of extra caution in the technique of power application during the turn. A short burst of partial power—that is aggravated by not having full forward control wheel or stick pressure—can combine with the tail wind to lift the tail and cause the airplane to end up on its nose or back.

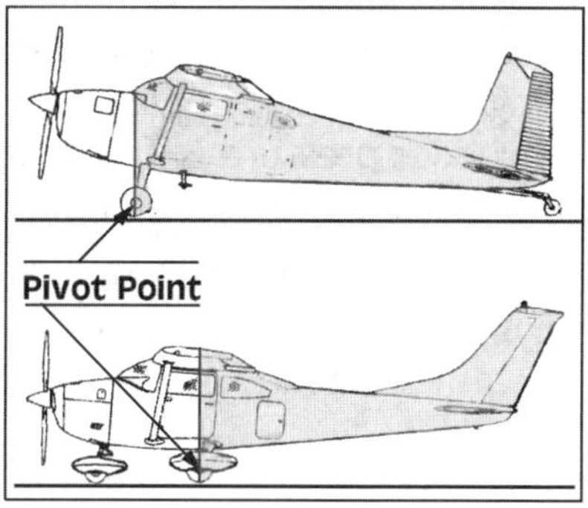

*Figure 8.* **Side area comparison.**

If you ever find the airplane tipping up (tail rising on its own), suck the stick into your gut, use a burst of power while releasing the brakes and accelerate out of the predicament. Then reduce the power while keeping the stick full aft. The amount of power necessary to get the tail down depends on the situation. Apply brakes as necessary to slow down with the airplane under control. It may be necessary to maintain power so the prop blast makes the elevator effective.

## Tail Rising On It's Own

- ✔ *Release brakes*
- ✔ *Add power to overcome the tailwind*
- ✔ *Stick full aft*
- ✔ *Slow application of brakes*

With the stick full aft, the high power provides a blast of air over the elevators increasing their effectiveness. Normally it is poor technique to apply brakes when power is applied, but not in this case.

## RUDDER USAGE

Test the brakes prior to moving the airplane very far from its tie-down position. If one or both brakes are inoperative, it is a simple matter to shut down the engine and stop the airplane.

Turns are made by depressing the rudder in the direction of the desired turn. Often, when the airplane is stopped or moving slowly, additional power and brake usage are required to initiate the turn. Prior to reaching the point where the turn is to be straightened out, apply opposite rudder to straighten the path. If the rudder pressure is neutralized too late, the airplane will continue beyond the desired path.

Experience allows the pilot to anticipate what the airplane is doing and to effect rudder changes before they are needed. By using this technique, less rudder is needed and the airplane is easier to control.

## VISIBILITY

Some tail wheel-type airplanes, sitting on the ground with a high angle of attack, block the pilot's view because of the engine cowling. Airplanes of this configuration require S-turns while taxiing to afford a view of the taxi surface and obstructions.

## CROSSWIND TAKEOFF

Crosswind takeoffs involve a different technique from the normal takeoff because the wind coming from the side tries to veer the airplane in the direction of the wind, while lifting the upwind wing. This produces a tendency to tip the airplane over or cause a swerve that leads to a ground loop.

*Figure 9.* **Crosswind takeoff**

The weathercock tendency is due to the airplane having more surface area (fuselage, vertical stabilizer and rudder) exposed to a crosswind aft of the main wheels than ahead of them. The taildragger, with its center of gravity located aft of the main wheels,

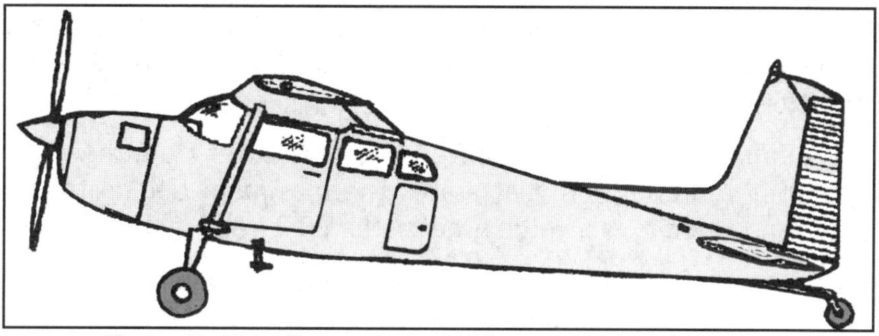

*Figure 10.* **Note the amount of side area behind the main-wheels pivot point on the Cessna 180.**

encounters a stronger weathercock tendency than a tricycle-wheel airplane with the center of gravity located forward of the main wheels.

A normal takeoff into the wind requires very little control input, other than rudder, by the pilot as the airplane tends to maintain a straight track; however, the crosswind takeoff requires special control input. Performed within the experience level of the pilot and the operating limitations of the airplane, the crosswind takeoff is relatively simple. The biggest problem and the most common cause of crosswind takeoff accidents is lifting off without sufficient airspeed to remain airborne. Then a touchdown while drifting causes a ground loop.

## WIND LIMITS

As a general rule, never attempt to taxi when the crosswind exceeds 50 percent of $V_{SO}$. Taxi slowly and with caution when the crosswind is 30 percent of $V_{SO}$.

Some manufacturers list a maximum demonstrated crosswind velocity greater than 0.2 $V_{SO}$. Their documentation may state the maximum demonstrated crosswind velocity is not a limiting factor (perhaps there wasn't a stronger wind available for their demonstration to the FAA), that implies that it is OK to takeoff or land in crosswinds stronger than that listed.

Regulations governing airplane design require an airplane be controllable in a 90-degree crosswind with a wind velocity of 0.2 $V_{SO}$. Operating the airplane within this limitation would be very restrictive. For example, the Cessna 170B has a $V_{SO}$ of 55 mph (47.8 knots). By regulation the airplane must be controllable at

0.2 $V_{SO}$ or 9.6 knots direct crosswind. If the pilot restricted his operations to this value the airplane could not be utilized to its full extent.

Insurance companies may have ideas that differ from the manufacturer's information that states the maximum demonstrated crosswind velocity is a value greater than 0.2 $V_{SO}$.

The manufacturer's demonstrated crosswind is not an operating limitation or authorization to operate at that value. It is usually listed in the POH under "normal procedures." It is a good idea to have an understanding of your insurance policy, in writing, concerning coverage during crosswind operations. Some insurance companies refuse to pay for damage if a regulation (their interpretation) is violated.

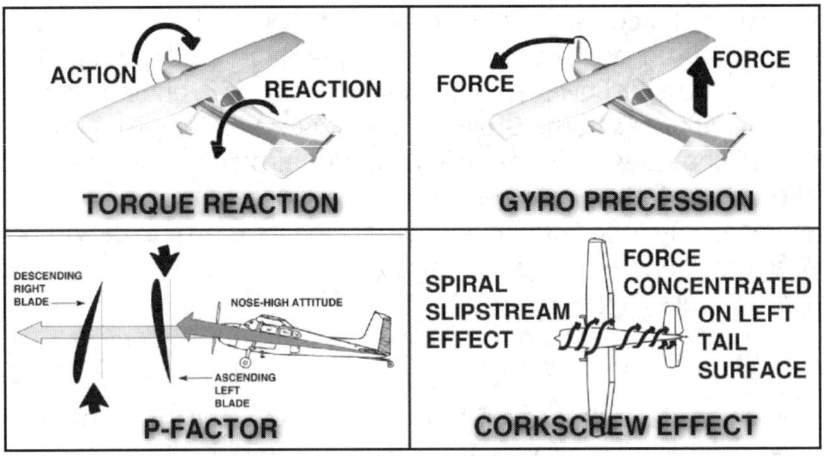


*Figure 12.* **The left turn forces acting on the airplane during takeoff. A crosswind from the left adds to these forces.**

## Crosswind-takeoff Technique

The crosswind takeoff is relatively simple for the pilot who understands the principles involved. Although it may, at times, be quite exciting. A crosswind, affecting the wings and other control surfaces, does not move in the same direction that the wheels roll along the ground. Therefore the takeoff technique results in part taxiing, part flying and the transition between the two.

While taxiing into the takeoff position, check the wind indicator for the trend of the crosswind. The action of the wind from the side does two things to the airplane. First, the wind tries to get under the upwind wing with a tendency to tip the airplane over; second, the airplane has a tendency to weather vane (streamline itself into the wind).

## Upsetting Wind Action

The upsetting action of the side wind is originally counteracted by holding the wings level or by trying to bank slightly into the wind by application of full aileron into the wind. If the wind is from the right, full right aileron deflection is held (the right wing aileron is up and the left wing aileron is down). A left crosswind requires left aileron pressure.

As speed increases during the takeoff roll, the upwind wing will develop more lift, causing it to "fly" or begin rising before the downwind wing. The use of aileron pressure is merely to keep the wings level.

The order of control effectiveness during acceleration is A-E-R, that is, aileron, elevator, and then rudder. With increasing speed during the takeoff roll, the ailerons become sufficiently effective to be able to maneuver the airplane about its longitudinal (roll) axis.

With greater effectiveness as the airplane accelerates, less aileron pressure is required. Until you develop a "feel" for the airplane, try relaxing one-half the aileron pressure when one-half the takeoff speed has been obtained. With experience, there is a gradual reduction of aileron pressure throughout the takeoff roll, rather than making mechanical movements. With a crosswind the ailerons are never relaxed to the neutral position until after liftoff.

## Skipping During Takeoff

If not enough aileron pressure is maintained, and the upwind wing rises, a "skipping" may result. This is a series of very small bounces, caused by the airplane attempting to fly and then settling back onto the runway while drifting some distance downwind. These bounces turn into side skipping where the imposed stress may cause the gear structure to fail or a ground loop to occur.

## Weather Vane Tendency

In addition to the use of aileron during the takeoff roll, it is necessary to use rudder and probably brake opposite the wind direction to counteract the weathercock tendency.

A weather vane is a thin plate of wood or metal that has often been shaped to resemble a rooster or arrow. Pilots are more familiar with the tetrahedron or wind tee. The weather vane pivots to indicate the direction of the wind.

The airplane's vertical stabilizer, rudder and side area make an excellent weather vane. The airplane tries to turn into the wind to become streamlined.

In a no-wind condition, the application of full power results in torque, p-factor, gyroscopic precession and the corkscrewing effect of the propeller slipstream. These turning forces yaw the airplane to the left.

During takeoff with a light crosswind from the right, the normal torque and p-factor may be sufficient to counteract the weathercock tendency. Conversely, with a left crosswind, it will add to the weathercock tendency, aggravating the situation. Regardless of the wind direction, use sufficient rudder to maintain directional control.

During the takeoff-ground roll the pilot must maintain a ground track parallel to the runway while compensating for the side forces trying to tip the airplane by lifting the upwind wing.

## Takeoff Technique

To perform the crosswind takeoff, begin the takeoff roll by using full aileron deflection into the wind. It may be necessary to use differential braking until the rudder becomes effective. If the crosswind is from the right, the airplane will try to turn right, into the wind. The left brake is used, but only until the rudder becomes effective enough to counteract the weathercock tendency. The use of the brake will lengthen the takeoff distance.

## Crosswind Liftoff

During a normal takeoff, instead of rotating at a certain distance or airspeed, slight back pressure is maintained on the elevator control until the aircraft is flown off the ground with little or no effort on the part of the pilot. In a crosswind takeoff, the aircraft is held on the ground with forward stick pressure until excess airspeed is obtained, then the pilot performs a definite and somewhat abrupt pull-off.

Extra speed—about 10 percent faster than normal—should be gained during the crosswind takeoff roll before attempting liftoff to prevent the airplane from settling back onto the runway in a crab. Make the pull-off definite when the increased flying speed (10 percent) is obtained. Then, immediately lower the nose to the normal climb attitude.

If the wind is gusty, add one-half the gust factor to the normal liftoff speed or fly a speed 10 percent faster than the normal lift-off speed, whichever is greater.

With sufficiently strong crosswind, it may be necessary to use aileron to prevent skipping. This may cause the downwind wing to rise and the downwind main wheel may lift off the runway first, with the remainder of the takeoff roll conducted on the upwind main wheel. This is acceptable and is preferable to side skipping.

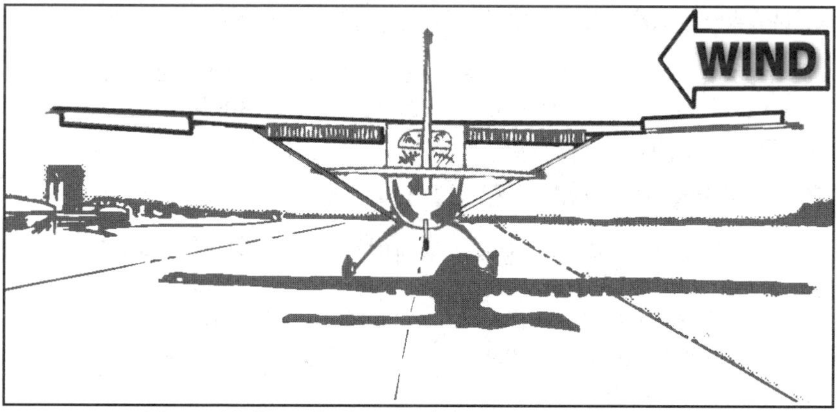


*Figure 13.* Use of ailerons during crosswind takeoff.

## Leaving the Ground

Once the aircraft is airborne with a positive rate of climb, smoothly apply coordinated control pressures to execute a turn into the wind sufficient to counteract drift. It is undesirable to correct for drift after liftoff by slipping because of the degradation of climb performance. Instead, establish a crab to compensate for the wind drift so the airplane tracks along the extended centerline of the runway. The remainder of the takeoff climb is the same as a normal takeoff and climb.

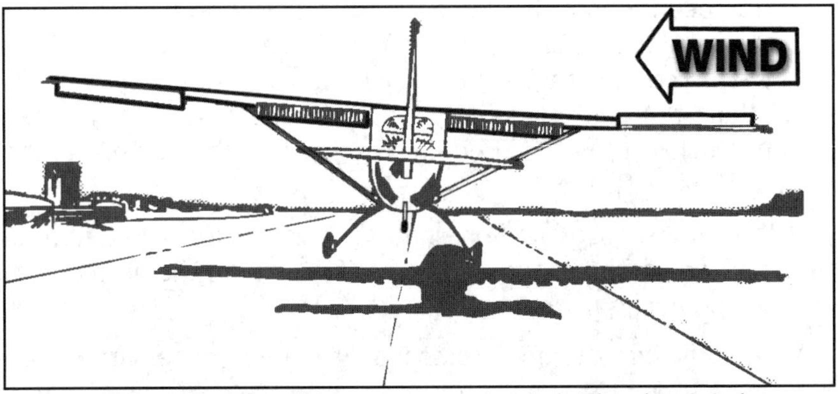


*Figure 14.* **Takeoff in a strong crosswind. The downwind wheel may lift off the runway before a positive rotation is made to compensate for the wind.**

## RUNWAY MARKINGS TO DETERMINE DISTANCE

The distance markings on an instrument runway may be used to make accurate estimates of distance for the angled-path takeoff and the curved-path takeoff.

## ESTIMATING RUNWAY LENGTH

Pilots who operate at back country strips, private strips or at an area they determine may be suitable to land on, for which there is no runway information, should make some determination of the runway length unless it is obviously long enough to operate without a problem.

In a no-wind condition the runway length may be determined before landing by flying the length of the runway and multiplying the time in seconds by the feet-per-second listed in the chart. If GPS is used an accurate determination of ground speed may be made.

If there is any wind and you try to average the ground speed by flying the runway in each direction, you will obtain an errone-

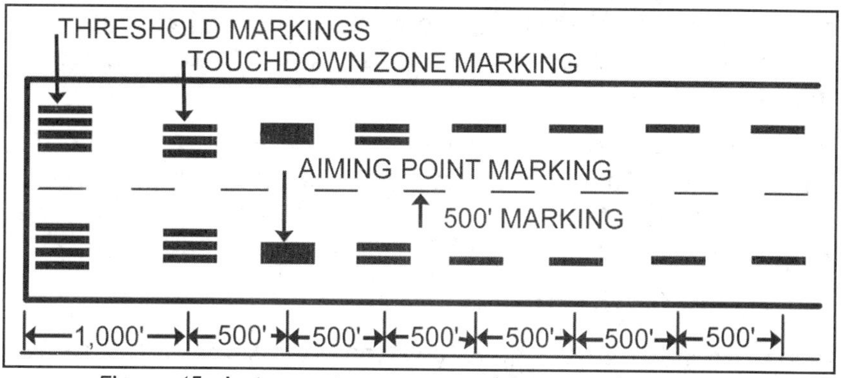


*Figure 15.* **Instrument runway markings can be used to measure distance. The number of stripes depicted by the threshold markings indicates the runway width (4 = 60', 6 = 75', 8 = 100', 12 = 150', 16 = 200')**

ous measurement. For example, assume a 20-knot wind blowing directly down the runway. With a 60-knot no-wind groundspeed, it will take 7.4 seconds to fly 1,000 feet at 80 knots. It will take 14.8 seconds to fly 1,000 feet in the opposite direction with a groundspeed of 40 knots. By adding together (22.2 seconds) the time and dividing by 2, the average time is 11.1 seconds. If you used an estimated ground speed of 60 knots, then 11.1 seconds is 666 feet for the runway length (11.1 seconds x 101.3 fps). This is an error on the conservative side; however, it lends no accuracy whatsoever, so don't waste your time.

## ANGLED-PATH TAKEOFF TECHNIQUE

Assume for this explanation that we are operating from runway 18. The wind is 270 degrees at 20 knots. The 20-knot crosswind is sufficient to present a challenge for the takeoff; however, the effect of the crosswind can be reduced by angling the takeoff path across the runway.

Line up on the downwind edge of the runway and point the nose at the upwind edge of the runway a distance equal to the no-wind take-

| Ground Speed - Knots | Feet per Second |
|---|---|
| 40 | 67.5 |
| 50 | 84.4 |
| 60 | 101.3 |
| 70 | 118.2 |
| 80 | 135.0 |
| 90 | 151.9 |
| 100 | 168.8 |

*Figure --.* **Estimating runway length based on time of overflight of the runway.**

off distance. For example, if the airplane requires 400 feet for a normal takeoff, point the nose 400 feet down the upwind edge of the runway.

If you can use runway markings to accurately determine distance, it is safe to use the exact takeoff distance performance for the angular takeoff because there will be a portion of the crosswind acting as a headwind which reduces the takeoff distance.

## CURVED-PATH TAKEOFF TECHNIQUE

When the crosswind is too strong to takeoff otherwise, the curved-path takeoff technique may be used. This procedure is for the experienced pilot, or an inexperienced pilot accompanied by a knowledgeable flight instructor. The technique involves flying from the downwind side of the runway toward the upwind side, with a turn to parallel the runway. The turn, being opposite to the wind, opposes the weathercock tendency. The curved path takeoff reduces the crosswind component.

### Runway Alignment

To see how the curved-path takeoff is used, assume operation from runway 18 with the wind from 240 degrees at 25 knots. The wind forms a 60-degree angle from the runway heading, giving a headwind component of 12.5 knots and a crosswind component of 21.7 knots.

Rather than lining up with the runway centerline, taxi to the downwind edge of the runway. Instead of aligning the longitudinal axis parallel to the runway length, point the nose about 10 to 20 degrees to the right of the runway centerline alignment, into the wind, toward the upwind edge of the runway.

This angle is variable and depends on the pilot's experience, the takeoff performance of the airplane and the runway surface. For a hard surface runway, it is determined by pointing the nose at an imaginary spot along the upwind edge of the runway. This imaginary spot will be located down the runway at a distance equal to two-thirds the takeoff distance required for the existing density altitude. If the nose is pointed farther down the runway, much of the advantage of using this technique is lost. If the nose is pointed less than this distance down the runway, the airplane may be unable to complete the required turn to parallel the runway once it arrives at the upwind edge of the runway.

For a soft field, such as grass or gravel, the angle cannot be as large as with the paved runway because the airplane may chatter or skid off the runway while trying to turn to parallel the upwind edge.

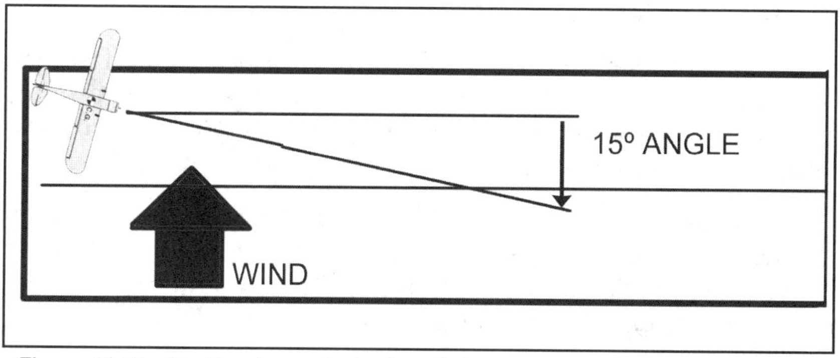


*Figure 16.* **Taxi to the downwind edge of the runway and align the airplane 15 degrees toward the upwind edge of the runway.**

## Reduction of Crosswind Component

We will assume the wind is a crosswind from the right for this explanation. Perform the angled-crosswind takeoff from the downwind edge facing the airplane toward the upwind edge of the runway. As the airplane approaches the upwind side of the runway, make a left turn so the longitudinal axis parallels the length of the runway.

There is a reduction of the velocity of the crosswind component realized by beginning the takeoff roll into the wind and reducing the wind angle during the initial portion of the takeoff. But, more important, the centrifugal force created by the left turn—that fights against the weathercock tendency of the wind keeps the upwind (right) wing down.

How much reduction in the velocity of the direct crosswind component is realized by angling into the wind?

Let's examine two different airplanes, one with a total ground roll takeoff distance of 400 feet, and the other with a total ground roll takeoff distance of 1,200 feet. Again we will depart from runway 18 with the wind 2425 (240 degrees at 25 knots). This imparts a direct crosswind of 21.7 knots (headwind of 12.5 knots).

Obviously, it will not be practical to compute the angle to turn into the wind, nor the exact crosswind component, before each takeoff. From the examples given, it should be possible to develop a rule of thumb for use under most conditions.

### EXAMPLE: Airplane One

For the airplane that requires 400 feet for the takeoff ground roll, the angle formed by pointing down the runway for two-thirds the takeoff distance—266.4 feet in this case— will vary with the width of the runway.

If the airplane made the takeoff, paralleling the runway (using a heading of 180 degrees), the direct crosswind component would be 21.65 knots. The curved-path takeoff allows a reduction in the crosswind component as follows:

With a runway 50-feet wide, the airplane would be turned 10.6 degrees into the wind with a direct-crosswind component of 15.86 knots.

The 75-foot-wide runway would allow the airplane to turn 15.7 degrees to the right with a resultant direct-crosswind component of 14.1 knots.

With a 100-foot wide runway, the angle would be 20.6 degrees to the right of the runway alignment of 180 degrees, resulting in a direct-crosswind component of 12.3 knots.

### EXAMPLE: Airplane Two

For the airplane requiring 1,200 feet for takeoff, the angle formed by pointing down the runway for two-thirds the takeoff distance, 800 feet, also varies with the width of the runway.

For a 50 foot-wide runway, the airplane is turned 5.6 degrees into the wind with a direct-crosswind component of 18.1 knots.

For the 75-foot wide runway, the airplane is turned 5.4 degrees into the wind with a direct-crosswind component of 17.6 knots.

With the 100 foot wide runway, the airplane is turned 7.1 degrees into the wind with the resultant direct crosswind component of 17 knots.

Without using the curved-path technique, the crosswind component would be 19.2 knots.

By angling into the wind the amount of crosswind affecting the takeoff is reduced. Again, it is the centrifugal force from the turn to parallel the runway that really helps in fighting the weathercock tendency and in keeping the upwind wing down.

To begin the takeoff, the elevator control is pulled back to keep the wheels solidly on the runway. The heading is maintained in a straight line while aiming at the predetermined point on the upwind side of the runway.

The pressure of the wind on the fuselage and empennage imposes a side strain on the landing gear during any crosswind takeoff. As long as a straight heading is maintained, this side strain is acceptable.

If the airplane is not held down firmly on the runway, it could skip sideways putting an unacceptable strain on the landing gear and its components.

If the wind is from the right side, the right wing will develop more lift than the left wing. It will tend to lift the upwind wheel off the ground. This occurs because the dihedral effect increases the angle of attack of the right wing and some of the left wing airflow is blanketed by the fuselage. To counteract this lifting action, move the control wheel (stick) to the right. The left aileron deflects down, increasing the lift on the downwind wing; the right aileron moves up, decreasing the lift and the drag on the upwind wing. As the speed increases and the ailerons become more and more effective, it is necessary to roll out some of the deflection. Use only enough aileron control to keep the wind from lifting the upwind wing, maintaining the wings level.

In a strong crosswind, that requires full-aileron deflection, the downwind wheel will liftoff first. The airplane is flown (on the ground) with the upwind wheel on the ground and the downwind wheel in the air, maintaining a straight ground track, until the proper airspeed for takeoff is obtained. The airplane is difficult to handle and you must force the airplane with forward elevator control to remain on the runway.

Avoid the temptation to haul the airplane off the ground before it will continue to fly without settling back to the runway. This procedure of maintaining only one wheel on the ground will prevent damaging side loads on the gear assembly.

### Common Errors

The most common cause of crosswind-takeoff accidents is making the liftoff without sufficient airspeed. Use approximately 10 percent greater airspeed than normal, or one-half the wind gust factor, whichever is greater. This prevents settling back to the runway in a crab.

- *Failure to maintain runway alignment on the takeoff roll.*
- *Climbing with one wing low after takeoff (side slip) instead of establishing a crab.*
- *Using too much forward elevator control pressure during the takeoff roll, resulting in loss of directional control.*

## CROSSWIND LANDING

Problems with crosswind landings can often be corrected by using a technique for simplified landings. The pilot incorporates attitude flying to establish the flare and landing attitude. From this point, the airplane mostly lands itself, with close monitoring by the pilot. (See: Simplified Landings)

Improper crosswind operations are responsible for the greatest number of landing accidents. There are several techniques for making a crosswind landing. Each results in aligning the wheels with the runway prior to touchdown to reduce side loads on the landing gear and associated structure.

Airline-type airplanes—with engine pods close to the ground on their swept wings—must use a crab/kick out method for their crosswind landing, to prevent ground contact of the engine pod, wing tip or inboard-flap section.

The crab/kick out method involves flying a crab to the flare and just before touchdown, the rudder is "kicked" to align the airplane with the runway. At the same time, opposite aileron from the rudder usage is used to maintain a wings level attitude. This requires practice with a competent instructor to prevent aircraft damage.

The wing down or side-slip procedure is considered the best method for the crosswind landing in light single-engine aircraft and will be discussed in four integral parts: approach, flare (or round out), touch down and after-landing roll.

### 1. Crosswind Approach

Crosswind landings are more difficult to perform than are crosswind takeoffs. This is mainly due to problems involved in maintaining precise control of the airplane while its speed is decreasing, combined with the additional factor of wind gusts.

It has been said that the approach is the key to a good landing. If this is true, then the approach for a crosswind landing becomes more critical.

What exactly is a good approach? The definition should state that it is an approach with proper altitude control throughout the descent combined with precise airspeed control, to arrive over a preselected point on the runway where a normal landing may be made.

A technique used to deal with the visual illusion of sloped runways, the 'spot method for landing,' is excellent for executing a good approach to a "flat" runway or a sloped runway. (See Spot Method)

### Establish Crab to Offset Drift

When turning onto final approach, establish sufficient crab into the wind to stop the drift. Some instructors advocate the pilot should immediately establish a side slip after tuning onto final. This is not always the best advice. As a general rule, the wind velocity decreases with decreasing altitude and shifts slightly counterclockwise. If the slip is begun after the turn from base to final, the airplane may not have enough rudder control to create a side slip sufficient to halt the drift. This may lead the pilot to feel he cannot land in the existing wind conditions. Yet, with the decreasing wind velocity and the wind shift closer to the ground, he may not have a problem.

Because the wind direction and velocity changes with altitude, use the crab for drift correction until one-quarter mile or less from the approach end of the runway. At that time, lower the upwind wing with aileron and simultaneously apply opposite rudder to maintain alignment of the nose (longitudinal axis) with the runway. It is likely the wind velocity will be changing; ailerons and rudder must change to accommodate the changing conditions of drift.

To properly accomplish the transition from crab to side slip, the aileron control (stick or wheel) is moved toward the crosswind. Because adverse aileron drag causes the nose to move away from the turn (the lowered aileron on the downwind wing increases lift; lift and drag are directly proportional), compensating rudder pressure is used.

### Forward Slip

The forward slip is a maneuver used to expose a large side profile of the airplane to the relative wind, creating drag, and resulting in a steep descent without an increase in airspeed.

To perform a forward slip, step on the rudder to move the nose sideways, and then apply aileron pressure to hold it. This results in the airplane moving forward through the air along the same flight path as before the slip, but with the nose pointed to the side. When coming out of a forward slip, relax both controls at the same time and the airplane will assume the same attitude as before the forward slip.

### Side Slip

The side slip is a maneuver used to fly the airplane laterally, an amount equal to wind drift, so it can touchdown on the runway with no side loading on the landing gear.

The exact same controls are used as with the forward slip; however, the ailerons are applied before the rudder in this case. The aileron application causes the airplane to fly sideways. The rudder application causes the longitudinal axis to remain parallel to the runway. When coming out of a side slip, relax the aileron control first, then the rudder. The airplane will assume the same attitude as before the side slip.

The amount of side slip, or compensation for the drift of the wind, is controlled by the amount of aileron deflection or wing down.

After turning onto final approach, maintain alignment with the extended runway centerline by flying into the wind an amount sufficient to counteract drift. This is known as flying a crab. The airplane, while crabbing, is in wings-level, coordinated flight.

The crab is maintained until descending below 200 feet above ground level. Because the wind changes direction and velocity as you descend, if the side slip is initiated too soon, the airplane may have insufficient rudder control to maintain the flight path in alignment with the runway. Below 200 feet, if the longitudinal axis cannot be maintained parallel to the runway, the crosswind is too strong for a safe landing that parallels the runway.

### Avoid the Mechanical Side Slip

Another problem in executing the crosswind landing with a side slip involves the pilot. Some pilots execute the side slip mechanically. Take a case where the crosswind is blowing from the right. The pilot lowers the right wing and applies left rudder. He maintains these control inputs throughout the approach and possible landing. The problem shows up when the wind shifts and blows from the left while the pilot maintains the slip to the right. The unsuspecting pilot, knowing in his mind the wind is from the right, maintains the slip to the right. He persists in the slip even when he has to stop the slip and fly back to the left in order to continue the slip to the right.

- *Do not fly mechanically.*
- *Use the ailerons, as necessary, to align the airplane laterally, correcting for drift.*
- *Use the rudder, as necessary, to align the airplane longitudinally, with the runway.*

Slips are often taught by rote, where the upwind wing is lowered and the opposite rudder is applied to maintain the longitudinal axis straight. Because of gusty winds that change direction without indication, it is preferable to use the ailerons to maneuver back and forth to stay near the runway centerline. The rudders are used to align the longitudinal axis parallel to the runway.

When rudder, opposite the aileron application, is applied to align the airplane's longitudinal axis with the runway, it will tend to level the wings. Additional aileron pressure is required to prevent this from happening. The lower the wing is, the more rudder deflection is required to maintain runway alignment. If the wind velocity is too great for landing, the rudder control will be deflected to its full limit (there will still be aileron control). When you run out of rudder control, it is time to make a go-around and land on a different runway, or use a different technique (See: Curved-Path Landing)

The most common error during the side slip is for the pilot to relax aileron pressure when the nose becomes misaligned with the runway. Apply sufficient aileron control to fly the airplane left or right to re-establish the desired lateral position. This is usually the runway centerline, unless the curved-path landing is being performed. Rudder pressure opposite the stick movement is used to maintain the desired alignment of the longitudinal axis due to the changing wind component and airplane speed. But, try not to fly mechanically. Apply whatever control is needed for lateral positioning (aileron) and longitudinal alignment (rudder). Do not depress the left rudder just because the right aileron is used.

During a side slip, the rate of descent will be greater. Additional power may be required during the approach. Airplanes equipped with a static vent on only one side of the fuselage may experience erroneous airspeed indications. A slip toward the static vent produces an erroneous low airspeed indication. It is necessary to fly an attitude (the relation of the nose of the airplane to the horizon) for airspeed control.

There may be a tendency to try to stretch the glide while in a slipping configuration. Do not raise the nose in an attempt to extend the glide distance. Maintain a constant approach speed (attitude) and apply power to lengthen the glide. (See: Spot Method)

## Gusty Wind Compensation

The final approach speed should be increased for gusty or strong crosswind conditions. The reason involves wind shear. An abrupt decrease of a gusty wind could cause the airplane to stall. One-half the difference between the steady wind speed and the peak gust speed is usually a sufficient amount to add to the normal approach speed.

## Flaps

Arguments arise when trying to decide about using flaps or not using flaps for a crosswind landing. You should use flaps as recommended by the manufacturer. There are two reasons for using flaps during the crosswind landing: there is greater aileron control and landing is at a slower speed.

Determine for yourself if there is more aileron control when flaps are used. While flying at a speed below the maximum flaps-extended speed, roll into a 20-degree bank. Maintain the bank constant with aileron pressure and extend one-half flaps. Notice the airplane rolls into a steeper bank without increasing the aileron deflection. Without changing the aileron pressure, add full flaps. The bank should increase again. Continue the same aileron position and retract the flaps. The airplane will roll out of the bank. This demonstration may provide evidence that the airplane has greater aileron control when flaps are used.

Using flaps reduces the touchdown speed. This makes it easier to stop an incipient turn or swerve. If a swerve occurs during a high-speed-landing roll, the centrifugal force or side loads can cause structural failure of the landing gear.

The slower the touchdown speed, the less centrifugal force is generated during a swerve. The relationship of speed to centrifugal force is often misunderstood.

Centrifugal force increases as the square of the speed when it begins. Doubling the airspeed means quadrupling the total centrifugal force. Even a small change in groundspeed at touchdown may affect the outcome of the landing. It is three times easier to land at 60 knots than 104 knots (104 knots squared is about three times 60 knots squared). It becomes obvious the actual touchdown speed should be the slowest possible, but controllable, airspeed.

The forces causing a ground loop will increase as the swerve increases. The first swerve develops centrifugal force and this, acting through the center of gravity, swerves the aircraft even more. If allowed to develop, the centrifugal force produced may become great enough to tip the aircraft until the wing strikes the ground.

Estimate the airplane's crosswind capabilities on short final before making a commitment to land. With the flaps at the desired setting, establish the slip correction. If you are not able to maintain runway-centerline alignment, you do not have enough control to handle the crosswind at slower speeds during the flare and landing roll. Discontinue the approach and try it again with a different flap setting to see if it makes a difference.

- *Flaps increase aileron control.*
- *Flaps reduce the touchdown speed.*
- *The centrifugal force is inversely proportional to the airspeed.*

## Curved-path Landing

The same basic procedure applied to the crosswind takeoff by angling across the runway, can be used for the crosswind landing. Approach the runway pointed slightly across, into the wind, and touch down near the downwind edge of the runway. Begin a gradual turn away from the upwind side of the runway to parallel the upwind edge. This will balance the upsetting force of the wind with centrifugal force and help keep the upwind wing down. Remember that centrifugal force increases relative to the square of the velocity where it starts, rather than in a linear manner, so try to keep the touchdown speed as slow as is safely possible.

## Twin-engine Taildraggers

The twin-engine taildraggers, such as the Beech 18 or DC-3, have the same directional-control problems at slow speeds as their single-engine cousins. During the crosswind-takeoff ground roll, advancing the power of the upwind engine in advance of, and in excess of, the downwind engine will help compensate for the weathercock tendency. Once sufficient speed has been obtained to make the rudders effective, the power output can be matched on both engines.

Some pilots claim—with conviction—the twin-engine taildragger, especially the Beech 18, is impossible to three-point land with normal loading. They claim the airplane must be loaded heavy in the rear to obtain a three-point attitude. While not experiencing this difficulty, it is interesting to note the normal and crosswind landing procedure in the FAA Approved Airplane Flight Manual for the various models of C-45, call for a wheel landing. Further, each landing calls for full flaps.

Some aircraft may experience directional control problems when using full flaps during a wheel landing. The flaps direct the airflow down and away from the empennage, blocking the airflow and making the rudder ineffective. This is especially precarious during the transition from tail-up to tail-down attitude during a wheel landing, where brakes are the only directional control device available to the pilot. If such is the case, use the manufacturer's recommendation for flap usage.

For crosswind landings, the twin-engine operator has the option of performing the crab/kick out, wing down or side slip, or using the differential power technique to maintain the longitudinal axis parallel to the runway centerline while stopping side drift.

The differential-power technique is not especially difficult. Once the power setting is established to maintain runway alignment, aileron is used to control the drift. The only problem is after the landing, where it becomes necessary to reduce the power evenly during the roll-out until the tail is firmly on the ground. Do not reduce both throttles to idle at the same time before the tail wheel is on the ground. If, for example, the throttles are split so that one engine is developing about eight or 10-inches manifold pressure more than the other, reduce the throttles slowly while maintaining the split power.

## 2. Crosswind Flare (Round Out)

The flare is begun at the same altitude as in a normal landing and in the same manner; however, the crosswind correction that was established during the approach is maintained. As the speed of the airplane slows, the controls will lose effectiveness in the following order:

- *Rudder first,*
- *Followed by elevator, and finally,*
- *the ailerons.*

With decreased control effectiveness, it becomes imperative to increase the control deflection of the ailerons and rudder, to maintain the same drift correction. Also, as the speed slows, the upwind wing will need to be held lower, as the effect of the crosswind becomes more influential.

Do not level the wings during the flare or touchdown as long as the crosswind is affecting the airplane. Otherwise, the aircraft will begin drifting and may touchdown in a crab or skip. If necessary, go around, but do not land while drifting or crabbing.

### 3. Crosswind Touchdown

The side slip causes the upwind wing to be lower. Touchdown is made on the upwind main wheel. As the forward momentum decreases after ground contact, the downwind main wheel will gradually settle onto the runway in a properly executed landing.

### Full-stall Landing

Crosswind landings in the full-stall attitude may be performed when concerned about controlling centrifugal force problems (a swerve leading to a ground loop). Once on the ground maintain full aft elevator control to hold the tail down and place more weight on the tail wheel. This results in more positive steering.

When a wheel landing is performed the tail should be lowered as soon as possible without generating enough lift to cause the main wheels to come off the ground.

The flaps should be retracted as soon as the airplane is on the ground to reduce lift. If the airplane is equipped with a Johnson Bar-type flap extension/retraction handle located between the seats, and you are unable to fully retract the flaps while looking outside at the runway, retract them as far as you can reach, then wait until stopped to retract them fully.

Exercise proper caution not to retract the landing gear if the airplane is so equipped. Thereafter, the landing is normal, except the aileron is held into the wind. As the airplane decelerates and the ailerons become less effective, more and more aileron is added until full travel of the control surface is obtained. I have caused myself more directional problems than the crosswind causes by using too much aileron too soon during the after-landing rollout.

### Wheel Landing

The majority of pilots known to the author prefer the wheel landing, without flaps, during crosswinds. The author prefers to feel it out for each landing, generally using the wheel landing, but occasionally using the three-point landing with flaps, except for large aircraft, during crosswinds.

During a normal landing, after touchdown, the power is reduced and the tail is allowed to settle. During the crosswind landing, the tail should be lowered, rather than being allowed to settle. This is accomplished with full back-pressure control wheel movement. The speed must be slow enough that the increased angle of attack will not cause the airplane to begin flying again.

During the period when the tail is lowered is the most critical part of the landing. The rudder will be ineffective, and if directional control is required, the use of brakes (and differential power on multiengine airplanes) is the only option.

An improperly executed wheel landing sets the pilot up for PIO (pilot induced oscillation) or porpoising. What happens is the pilot makes an acceptable approach and flare. At the moment of touchdown the pilot is slow in adding the forward pressure necessary to maintain the wheels solidly on the ground. This occurs because the pilot did not expect the main wheels to touch (usually because of focusing his vision too close to the airplane), or he is hesitant about using forward pressure, having an unconscious fear of striking the propeller.

In either case the pilot then, after the airplane skips, applies forward pressure. He now sees the airplane approaching the runway with a nose down attitude and pulls back, but not enough to stop the ensuing bounce. This scenario of forward-and-aft pumping on the elevator control may be repeated many times before the pilot decides to:

- *perform the go-around procedure.*
- *add full-aft-elevator control at a rate that does not cause a balloon until the airplane is firmly on the ground.*
- *or, the pilot suffers a propeller strike.*

Often, in learning to execute the wheel landing, it is helpful for the student to trim the airplane with forward or nose-down elevator control. Then, when the wheels make contact, relax the back pressure and the airplane will stay on the ground.

## 4. CROSSWIND AFTER-LANDING ROLL

With the exception of wind-shear conditions, the only wind that acts on the aircraft in flight is the wind created by the aircraft's movement through the air. This is called the relative wind, that is, it is relative to the flight path, being equal, but in an opposite direction. During a side slip the relative wind will be slightly to the lowered-wing side of the airplane.

When the airplane is on the ground, it becomes influenced by two winds: The relative wind due to the forward movement of the airplane, and the movement of the air mass (wind). Somewhere between these two component winds there evolves a resultant wind.

The angle of the resultant wind to the aircraft depends on the velocity and direction of both the relative wind and the crosswind. As the aircraft slows during the after-landing roll, the resultant wind moves more broadside (toward the crosswind), because the relative wind is diminishing, while the crosswind remains constant.

During the after-landing roll, particular attention must be given to maintaining directional control with the rudder, while keeping the upwind wing from rising by use of aileron.

The airplane has a greater side area behind the main landing gear, considered the fulcrum point, than in front of the landing gear. With the main wheels acting as a pivot point and a greater profile behind the main wheels, the airplane naturally wants to turn or weather vane into the wind.

Because of the increasing sideward angle of the resultant wind as the aircraft's forward speed decreases, it is vital that increasing aileron be applied. But, it is equally as important in the tail dragger, that the aileron is not used to the extent the desired flight attitude is changed.

Use enough aileron deflection to prevent the crosswind from lifting the wing, but not so much that the wing will start down. As the airplane decelerates further, it may become necessary to hold full aileron deflection to keep the windward wing down. More rudder pressure will also be necessary to prevent the weather vane. The use of downwind brake pressure may also be required.

## CROSSWIND LANDING ERRORS

### Failure to Correct for Drift

The most common error in crosswind landings is failure to correct for drift at the moment of touchdown. Either too much to too little drift correction, or a touchdown in a crab, will impose potentially damaging side loads on the landing gear. Misalignment of the wheels to the ground path will also result in erratic directional control, possibly leading to a ground loop. The best correction for ground loops is prevention.

### Ground Loop

"Fly the thing 'til it's tied down," is sage advice for the conventional gear pilot.

### Excessive Aileron Aggravates

Pilots often associate the ground loop with the after-landing roll. The ground loop is an uncontrolled turn during ground operation. It may happen while landing, during taxi or during the takeoff. It is not always caused by drift or the weather vane tendency. Careless or improper use of rudder, uneven ground or a soft spot that retards one wheel may also cause a swerve. The initial swerve makes the aircraft ready for the ground loop. When the swerve occurs, centrifugal force causes the airplane to veer even more. Centrifugal force increases as the square of the speed where it begins, so the faster the airplane is going, the worse the centrifugal force.

If a ground loop begins at slow speed, reduce the power and use the controls to straighten the plane. If the speed is fast enough for the rudder to be effective, use power to counteract the centrifugal force developed during the swerve, and coordinated use of the rudder, brake and ailerons. This is the only remedy for stopping the ground loop at faster speeds.

If power and control usage do not stop the ground loop or allow a go-around, it is best to keep flying the airplane until its energy is expended, attempting to prevent the outside wing from dragging on the ground. Violent application of the brakes, especially if the wing tip contacts the ground, may result in the airplane flipping over onto its back.

### Runway Overrun

Another error occurs with a high-speed approach and touchdown resulting in overrun of the runway.

### Excessive Wind

Another accident producer occurs when the pilot tries to land when the wind is too strong. If, during the approach below 200-feet agl, full rudder is applied and drift still occurs, the wind is too strong for a safe landing. Make an immediate go around.

### Maximum Crosswind Component

FARs require all airplanes type certificated since 1962, to have safe ground-handling characteristics with a direct crosswind component equal to 0.2 $V_{SO}$. Thus an airplane having a stall speed in landing configuration (flaps extended) of 50 knots, must have no uncontrollable ground looping or weather vane tendencies with a 90-degree crosswind component of 10 knots (0.2 x 50).

Most airplanes manufactured since 1962 have a demonstrated crosswind velocity that is greater than the safe handling characteristics required by FARs. This is a velocity that has actually been demonstrated during certification tests where the airplane displayed adequate control during takeoff and landing. Although the manufacturer explains the value is not an aerodynamic limit for the airplane and is not to be considered limiting; check with your insurance carrier to determine its policy should a claim be submitted when a landing is attempted in a crosswind that exceeds the demonstrated value.

General rules of thumb for crosswind operation:

- *Never land when the 90-degree crosswind exceeds 20-percent $V_{SO}$.*
- *Never land when the 45-degree crosswind component exceeds 30-percent $V_{SO}$.*
- *Never land when the head wind component exceeds 60-percent $V_{SO}$.*

## Crosswind Component Chart

The value of the wind blowing at an angle to the runway can be converted to determine the direct-crosswind component with a crosswind velocity chart or circular computer.

Determine the relative angle of the wind (RAW), by subtracting the runway direction from the wind direction.

EXAMPLE: 40 knot wind, 30-degree angle

- *Plot the 30-degree angle between the wind direction and runway direction.*
- *Move to the 40 knot wind velocity arc.*
- *Move horizontally left to determine 35-knot head wind component.*
- *Move vertically downward to determine 20-knots crosswind component.*
- *Plot the RAW onto the curved portion of the crosswind component chart at the point where the RAW and wind velocity intersect.*
- *Move vertically downward from the RAW and wind velocity point, to determine the crosswind component. If you want to know the head wind component, move horizontally to the left from the RAW and wind velocity point, to read the head wind component.*

EXAMPLE: If landing on runway 18 with the wind 240 degrees at 23 knots, the RAW is 240 minus 180 = 60 degrees. Plot this position and move vertically downward to read a 20-knot crosswind component. (*See: Figure 17*)

## GO/NO-GO DECISION

An experienced pilot may find the following figures (recommended by the FAA) to be conservative. In some operations, they are too restrictive. For the new taildragger pilot, the figures provide a margin of safety to allow one to build experience without encountering dangerous situations.

✔ *Determine the maximum 90-degree crosswind for your airplane. If this is not listed in the owner's manual or pilot's operating handbook, use 20 percent times the stall speed. Place a dot on the "crosswind component computer"*

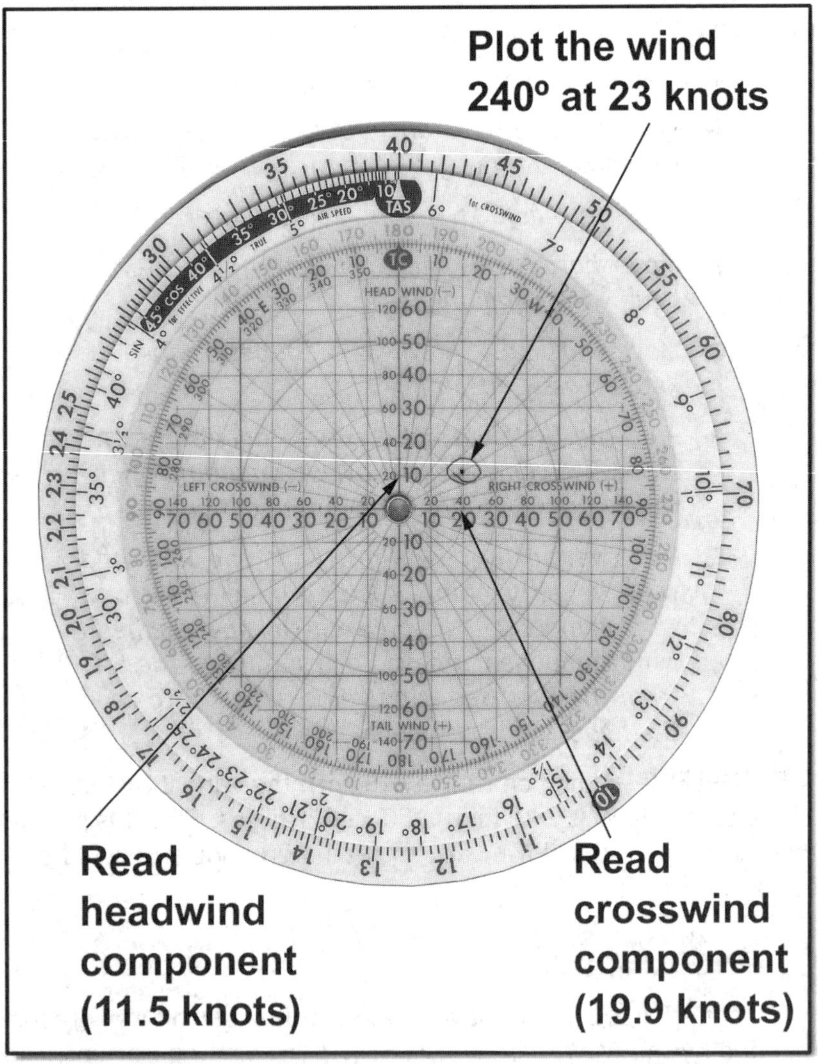


*Figure 17.* Align 240° (wind direction) with the TC index. Measure up 23 (knots) and make a mark. Rotate inner scale so 180° (runway) is under the TC index. From the mark, move horizontally left to read the headwind component. Move vertically downward to read the crosswind component.

*on the bottom horizontal line labeled "cross-wind component."*

- ✔ *Determine the maximum 45-degree crosswind for your plane. It is suggested you use 30 percent times the stall speed. Place a dot on the 45-degree line at this value.*
- ✔ *Determine the maximum head wind for your airplane. It is suggested you use 60 percent times the stall speed. Place a dot on the "head wind component" line of the chart at this value.*
- ✔ *Connect these dots with a red line.*

Any values determined to be left of the red line are GO.

Any value determined to be right of the red line are NO-GO wind directions and velocities.

Circular (CR) Computer Determination of Crosswind Component

- ✔ *Place the wind velocity above the TAS index on the backside of a circular computer.*
- ✔ *Read the crosswind component above the RAW (determine the relative angle of the wind by subtracting the runway direction from the wind direction). This is the middle scale. The white area of this scale represents the SINE. The black area represents the COSINE.*

If desired, the head wind or tail wind component can be read above the cosine of the RAW (Since the COS scale only goes from 0° to 45°, the Cosine = 90 minus RAW, or 30).

EXAMPLE: Landing runway 18. Wind 240 degrees at 30 knots. The RAW is 240 minus 180 = 60 degrees.

- ✔ *Place the wind velocity, 30 knots, above the TAS index.*
- ✔ *Above 30 (black scale) whenever the value exceeds 45 degrees, it is necessary to subtract*

*that value from 90 to obtain the sine, in this case, 90 minus 60 = 30.*

✔ *Read 26-knots crosswind component.*

The head wind component (white scale) is read above the 30 mark, as 15 knots.

*NOTE: This example borders on the extreme where the relative angle of the wind exceeds 45 degrees. Normally the relative angle of the wind will be less, and the SIN and COS values can be read directly from the circular computer scale.*

# CHAPTER 10

# Tie-down

## INTRODUCTION

Without some specialized (but simple) knowledge, one's investment in a taildragger can be lost in a moment's time. The accepted procedures for tying down the taildragger are presented here to help preserve these airplanes.

## REMOTE STRIPS

Many small airstrips do not have adequate tie-down facilities. When operating near mountain areas, strong, gusty winds may arise without warning or apparent cause. It is prudent to purchase or make a tie-down kit to carry with you. Screw-type anchors, like those you can purchase at a pet store, offer greater security than pitons.

*Figure 1.* **Carry a tie-down kit in the airplane.**

If you are caught at an "out back" airstrip without a tie-down kit, use an emergency procedure that will enable the airplane to withstand an unbelievable amount of wind. Dig holes in front of the main wheels and roll the airplane's wheels into the holes. The change in reducing the angle of attack minimizes the effect of the wind. Do not dig the holes deep enough to damage the gear or fairings.

*Figure* 2. Given the choice, rope is better than chain for the tie-down.

The technique of digging a hole for the main gear could damage the fairings if you aren't careful.

If the airplane is left untied for any length of time, spoilers may be made from strips of wood of any length that measure one or two inches square. These are tied parallel to and on top of the first third of the wing. At least 50 percent of the wing span should be covered. The wing surface should be padded to prevent scratching (articles of clothing will do if nothing else is available). If wood strips are not available, use your imagination. Tires, sand bags, flour sacks, large sugar sacks, flight bags, survival kits, sleeping bags, easily removable pilot and passenger seats, and the like, can be used to fashion spoilers.

Figure 3. Tie down the airplane even if you will be gone for a short time period. Here the prop blast (seen by the dust) is more detrimental than the wind (windsock on right). Johnson Creek, Idaho

Another method used by many taildragger pilots when the airplane is left unattended for short periods of time, is to lock one brake only, and let the airplane weathercock into the wind. It is usually better to set the upwind brake. Be sure to give the airplane "breathing" room away from other airplanes or obstacles while it maneuvers around the locked wheel. For airplanes with toe brakes, for the brake that is to remain unlocked, press the lower part of the rudder pedal without applying the brake. The other foot presses the toe brake of the other pedal. Apply the parking brake mechanism. For heel-brake-equipped airplanes, apply one brake and set the parking brake.

Many of my favorite fishing areas allow landing on roads, fields or unimproved air strips. Livestock in the area, mainly cattle, require precautions to guard and protect the airplane. Some strips have a fenced area—to keep cattle out, not in—where the airplane can be secured. If you fail to guard the airplane, upon your return you may find the cattle have eaten portions of a fabric plane or rubbed against the horizontal stabilizer/elevator or struts of metal planes, causing damage.

### Improved Airports

Back in the '60s, a good used airplane could be purchased for $1,500. There was a great deal of complacency among the pilots of some of these airplanes. They would fly in somewhere and walk away from their airplane, leaving it untied for hours at a time. I figured this must be due to the cheap cost of the plane, but I was wrong. Airplanes are no longer as affordable, yet the same complacency is often prevalent.

A taildragger must be tied down when left unattended for more than a few minutes. It should be tied down, even if you remain at the airport, when it is left unattended.

## PREVENTING DAMAGE

Sometimes damage to airplanes is caused by improper tie-down sense, rather than the wind itself. Many experts advocate that chain tie-downs are the best, and the majority of fixed base operators use chains.

The problem is, chain material is not elastic. There is absolutely no give. This causes problems if the chain is too tight or too loose. In a high wind condition, all the wind force is transmitted directly to the strut and airframe (including the wing spars). Ropes, with their elasticity, help minimize this shock.

The tricycle-gear pilot can often get by with jumping out of the plane and securing the wing tie-downs and ignoring the tail tie-down.

The taildragger pilot, if wanting to eliminate a tie-down, should secure the tail and one wing. The tail tie-down will prevent the airplane from 'walking' or flipping over on its nose.

## TIE-DOWN SENSE

When considering tie-down materials, such as chain or rope, do not fail to think about and inspect the tie-down anchor. Both the tie-down anchor and tie-down rope (or chain) should withstand about 3,000 pounds of pulling force for the single-engine taildragger. It is recommended the strength be 4,000 pounds for large single-engine and twin-engine taildraggers.

### Chains

To deal with chain tie-downs, one might carry some short lengths of rope (3 to 4 feet) to attach to the chain and to the airplane tie-down ring to provide the elasticity required to eliminate damage.

If chains are the only tie-down available, place one loop of the chain through the tie-down ring and secure with the s-ring. When the wind rocks the airplane, there is a possibility this s-ring can disengage. To eliminate this possibility, some pilots will place a second chain link through the first on the other side of the tie-down ring, and then lock it with the s-ring.

### Ropes

**Manila rope** *is manufactured from Manila hemp. As an organic material it is susceptible to mildew and rot. It has a tendency to stretch when wet, then shrink when drying. In new condition, 1/2-inch manila has a tensile strength of 2,650 pounds; 5/8-inch manila has a tensile strength of 4,400 pounds. Sisal (Mexican agave leaves yielding a strong white fiber) and hemp (Asian herb fibers from cannabis) are not recommended.*

**Nylon rope** *of 3/8-inch diameter displays a minimum tensile strength of 3,400 pounds and 7/16-inch diameter gives 4,800 pounds tensile strength.*

**Twisted Dacron** *rope is stronger than braided, with 3/8-inch diameter giving 3,120-pounds tensile strength, and 7/16-inch yields 4,500 pounds.*

**Twisted-yellow Polypropylene** *returns more tensile strength than braided. With 3/8-inch, it gives*

*2,750-pounds-holding power, while 1/2-inch diameter provides 4,200-pounds-tensile strength.*

Any of these ropes are suitable for tie-down purposes; however, the synthetic ropes of the same size are stronger and will withstand the elements for a greater period of time than organic ropes. They will be less susceptible to "bruising" in the tie-down ring areas that provide direct metal-to-rope contact.

Because of the difficulty of securing a non-slipping knot in some polypropylene ropes, you may find the nylon or twisted Dacron to be the best choice.

## TIE-DOWN PROCEDURE

Where the tie-down rope or chain is secured to a long cable, determine the cable is secured at various points along its length, not just at each end. Ropes and chains attached to a cable can be slid back and forth. A strong side wind could shove several airplanes together if the cable is not anchored at regular intervals along its length.

*Figure 6.* Locking half-hitch. Begin by forming a loop and pulling the rope through.

If the tie-down is secured to the ground, inspect the anchor and attachment. If they look flimsy, check with the fixed base operator to determine the tensile strength.

REMEMBER

The best knot for securing a rope to the ground anchor is the bowline. It will not slip or slide, and it is easy to untie regardless of how much pull has been placed against it.

*Figure 7.* Pull the rope through the loop again.

Having selected the proper rope with tensile strength above and beyond that recommended will not do much good if the knot used to secure the airplane is lacking. Once the rope is passed through the wing tie-down ring, it may be secured with a bowline or double-half hitch. Most experienced pilots use the double-half hitch (locking-half hitch). Move down four to six inches and tie another double half hitch. My grandpa always told me, “A double-half hitch will hold the devil.”

The rope is secured; pulling it tight so there is between one and two inches of movement to the side. This sideward movement does not result in a dangling rope. The rope has some tension on it, sufficient to keep it tight, yet loose enough to allow the airplane to rock back and forth on the gear, relieving some of the pressure transmitted by the force of the wind.

*Figure 8.* **With the rope held to the side, pull tight and it locks.**

The tail wheel rope is tied to the ring on the tail wheel. Without a tail-wheel ring, the rope is wrapped around the spring or strut close to and above the tail wheel.

Airplanes without wing tie-down attach points are susceptible to damage if tied improperly. Obtain the manufacturer’s recommendations for securing an airplane without tie-down fittings. If the manufacturer is no longer in business, have your maintenance personnel check maintenance and service manuals.

*Figure 9.* **The completed locking half-hitch with slack tied up. A double half-hitch will hold the devil.**

## Chocks

The use of chocks on the main wheels of taildraggers should be a matter of habit, even when the airplane is secured with tie-downs.

## Brakes

The obvious time to tie-down an airplane is directly after a flight. Even with minimal use, the brakes become hot. Do not apply the parking brake immediately after use, especially with hot ambient temperatures, unless the airplane is equipped with a thermal expansion valve or chamber. This valve allows expanding brake fluid to escape, rather than blow out the o-rings and seals in the master cylinder or wheel cylinders.

## Control-wheel (Stick) Locks

Some airplanes have a provision for the installation of a mechanical lock inside the cockpit that secures the ailerons and elevator, or ailerons, elevator and rudder. These control locks or gust locks prevent the flying surfaces from banging against their stops.

If an airplane does not have a mechanical lock, use the seat belt to secure the stick or wheel. Anticipating a strong blow, use both seat belts to secure both sticks (control wheels).

For an airplane equipped with a control stick, pull the stick aft and wrap the seat belt around the shaft two times. Secure the seat belt clips and pull tight.

With those airplanes equipped with control wheels, pull the control wheel aft and turn it in the direction of the prevailing wind (destroy lift on the upwind wing, increase lift on the downwind wing). If the airplane is tied-down with a tail wind, turn the control wheel so the upwind aileron is down. Run the seat belt through the yoke and connect it to the other end of the belt, then pull tight.

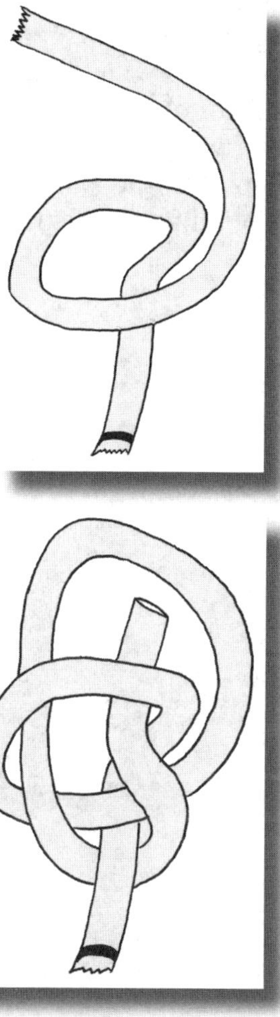

*Figure 7.* **Bowline knot. Begin by forming a loop over the main length of rope. Bring the loose end up through the loop, around the main length and back down through the loop.**

A gust lock for the rudder is easily made and your mechanic will be glad to provide ideas and suggestions on the proper way to fabricate such a lock for your particular make and model aircraft. He will probably admonish you to tie a large colored strip to the gust lock to remind you to remove it before flight.

Some airplanes are equipped with flaps that are held in the retracted position by springs rather than with a locking mechanism.

Airplanes parked so the wind approaches from the rear can damage the flaps. A simple control lock can be fabricated for the flaps. Cut two 5" x 6" strips of 1/4" plywood. Glue outdoor carpet to one side of each piece. Drill a hole in the center and insert a threaded bolt with a wing nut. The bolt is placed in the crack between the flaps and the wing root. Consider adding a strip of bright color to remind you to remove the locks before flight.

# CHAPTER 11

# Cold Weather Operations

## PILOT PREPARATION

Many aviation publications have printed articles that encourage the average general aviation pilot to keep flying during the winter months. Some of the articles contain practical advice on operating procedures, rules-of-thumb and precautions. Others may cause you to get into trouble with their advice. Analyze what the article professes as the proper procedure and if you have questions, ask your instructor.

*Figure 1.*

The author has never felt that it is proper to try to convince someone, one way or the other, about cold-weather flight operations (or night-single engine operations over the mountains or IFR-single-engine operations over the mountains). Whether you do it or not is something for you to decide. If you want to or need to fly during cold weather, you need to do it properly.

Most of us are not comfortable in expanding our comfort zone through trial and error. We don't step outside of a circle that encompasses the known without reading books and engaging flight instructors to provide us with the additional knowledge for a safe operation. That is, the knowledge that we can do something without suffering dire consequences.

By exposing you to information about cold weather and winter flying operations, the possibility of being exposed to something unknown is reduced.

The pilot must be equipped, in addition to knowledge, with the proper attire for the flight operation. This includes cold weather clothing and a survival kit.

## AIRCRAFT PREPARATION

All mechanical equipment, including airplanes and their components, are designed for operation within certain temperature ranges. Exceeding the temperature range of the airplane or equipment may create premature failures.

*Figure 2.*

You should determine the temperature ranges for operating your airplane. Only the manufacturer can make predictions of the product's performance. If the information about these temperature extremes is not readily available in the airplane flight manual, contact the manufacturer for an outline of the precautions to be taken in cold weather to prevent failures.

Without specific information about your airplane, seek the advice of competent operators and mechanics permanently situated in the location where you wish to operate. They can judge the requirements and limitations for the peculiarities of their area.

### Insulation (Reciprocating Engine)

If the temperature is very cold, insulation, called lagging, should be installed on all oil lines, oil pressure lines, and oil tanks. Without lagging, the oil may congeal, leading to oil pump cavitation and engine failure.

The insulation must be of a fireproof material. Installation by an experienced A&P mechanic is a necessity, since the mounting of insulation depends on how far the lines and tanks are from the heat source, the rate of oil flow, and the expected range of temperatures.

## Baffling and Covers

Some manufacturers provide baffles, winter front covers, and oil cooler covers, as optional equipment items for winter operations.

If the aircraft manufacturer does not provide baffling and covers, FAA approval is necessary for installation of such items. For example, it is not legal to cover a portion of the oil cooler with duct tape.

Installation of baffles suggests a cylinder head temperature gauge should be used, particularly if you operate in areas with wide temperature variations.

## Engine Oil

Oil is the lifeblood of the aircraft engine and its importance cannot be over stressed. It must lubricate, cool, seal, and clean the engine. You can't blindly pour any motor oil into your engine. The flight manual (or perhaps the filler opening) will tell you to add only oils of a particular type and grade. An older airplane may not have the new multi-viscosity oils listed, but Phillips Petroleum and Shell have their multi-viscosity oils approved by the FAA and they meet the requirements of Avco Lycoming Specification No. 301F and Teledyne Continental Motors specification MHS-24B. You can safely use them in your Lycoming or Continental engine.

## Oil Viscosity

To identify oils, the petroleum industry devised a classification system for engine lubricants. This system is based on grades. Viscosity, or thickness, determines the grade. Viscosities have been assigned identifying SAE (Society of Automotive Engineers) numbers. Classification of commercial aviation oils is by symbols that approximate SAE numbers.

The viscosity of the oil is very important in cold weather operations. If the oil is too thick, it will not lubricate the engine when cold. If too thin, it will break down at high temperatures. Until recently it has been necessary for pilots to switch back and forth between "summer" and "winter" grades. And, some pilots changed oil whenever they operated in different parts of the country. Even for someone not bothered by seasonal-oil changes, multi-viscosity oil is good for the engine.

| COMMERCIAL AVIATION NUMBER | COMMERCIAL SAE NUMBER | AN SPECIFICATION NUMBER |
|---|---|---|
| 65 | 30 | 1065 |
| 80 | 40 | 1080 |
| 100 | 50 | 1100 |
| 120 | 60 | 1120 |
| 140 | 70 | — |
| OIL CLASSIFICATION | | |

*Figure 3.* **Oil Classification.**

## Multi-viscosity Oil

Multi-viscosity oils will prolong the engine life as the airplane climbs or descends through temperature extremes. They add to engine life because they are thin enough to flow when cold and thick enough to be effective when hot. They save wear and tear not only on the engine, but also on accessories such at the battery, starter, oil pump, oil cooler and propeller governor.

The multi-viscosity oils are compounded with a complete package of anti-scuff, anti-wear, anti-oxidation, anti-acid, anti-corrosion, and other additives. Some of the anti-wear additives will make scored and galled surfaces inside the engine heal themselves. That's something the best straight-grade oil could never do.

Claims that multi-viscosity oils will permit longer intervals between oil changes should be disregarded. Most people think that as long as the dipstick is not a black molasses consistency, the oil is still doing its job. But, since there are many by-products an oil must clean, such as rust, corrosion, acids, and varnishes, stick to your regular oil change schedule.

## Pour Point

The pour point of the oil is important in cold weather operations. Once the temperature drops below the pour point of the oil, the oil thickens and solidifies rapidly with further temperature decreases. The pour point is the lowest temperature that oil will flow without any disturbing force.

| COMMERCIAL AVIATION NUMBER | POUR POINT |
|---|---|
| 100 | +10°F |
| 65 | 0°F |
| 20w-50 Phillips X-C | –25°F |
| 15w-50 Shell | –30°F |
| POUR-POINT OF OIL | |

*Figure 4.* Pour-point of oil.

## Fluidity

If the fluidity of the oil, based on the pour test and viscosity, is good, it will provide immediate circulation after starting the engine. Straight-grade oils cannot compare to the multi-viscosity oils in this respect.

## Oil Breather Lines

Moisture forms whenever combustion occurs. Most of the moisture is scavenged out the exhaust. Some moisture, however, finds its way into the crankcase. When these crankcase vapors cool, they condense in the crankcase breather line. If the ambient temperature is at freezing or below, the breather line may freeze closed. When this occurs, pressure builds up in the crankcase that might cause the oil filler cap to blow off or an oil seal to rupture, causing oil to pump "overboard."

Special care is required during the preflight inspection to assure the breather system is free of ice. Some pilots have their mechanics drill a small hole in the breather line where it comes out of the crankcase, or cut a slit, if the line is a hose, to relieve the pressure if it freezes. This hole may result in a dirty engine compartment; whereas, if a punch is used to indent an area above the hole or concave the hole surface, the pipe will breathe without leaking oil from the hole. The hole or slit should be located where its pressure relief effect will not direct oil toward an exhaust pipe.

## Cabin Heater

Checking the cabin heater as part of the preflight is in order for all pilots when temperatures approach freezing at flight altitudes. Many aircraft are equipped with cabin heater shrouds that enclose the muffler or partitions of the exhaust system. A thorough inspection is necessary to eliminate the possibility of carbon monoxide entering the cockpit.

Without adequate cabin heat, sub-zero air can reduce cockpit visibility to zero in a matter of minutes. Moisture from the pilot's breath may be sufficient to fog or even ice windows and windshields. Pilots caught in such a predicament have had to improvise frantically with plastic credit cards or control wheel locks as ice scrapers, while striving to keep the airplane level and on course.

Attempting to continue a flight once a failure has been discovered in the heating system is foolhardy. If the failure becomes known after the airplane is aloft, the pilot on an IFR flight should make his condition known to ATC, and, if necessary, request vectors to the nearest suitable airfield. Air vents may be opened, if feasible, to prevent the accumulation of moisture in the unheated cockpit. This will prevent the moisture from condensing and freezing on the windshield. Unpleasant as exposure to sub-zero air may be, it does not compare to the discomfort of trying to land with zero visibility.

Lack of heat can be pilot incapacitating to the extent that one cannot change radio frequencies or perform a simple task like picking up a pencil. It also results in the pilot exhibiting impaired judgment.

Although the likelihood of cabin heat failure occurring in flight is small, it does happen. One of the most common reasons for gas heater failure is not mechanical failure, but rather, a frozen fuel supply line. Approved anti-ice additives for the fuel helps, but no pilot should attempt to fly through wintry skies without the presence of cold-weather gear within reach of the cockpit.

## BATTERIES

Wet-cell and dry-cell batteries will require special consideration during cold weather for acceptable performance.

### Wet-Cell Battery

If an airplane is kept outside during freezing temperatures, the wet-cell-storage battery (similar to an automobile battery) must be kept fully charged to prevent loss of power and freezing.

Neglecting to maintain the correct fluid level is the most common reason for the aircraft battery to freeze. Also, inactivity may cause the battery to become drained of its electrical charge, allowing it to freeze.

If a battery becomes frozen, it may crack and leak sulfuric acid. If it doesn't break from the freezing, when subjected to a high-charge rate, it may explode.

If the airplane is parked outside for prolonged periods of time during cold weather and the battery is not fully charged and no provision exists for charging it, remove the battery and store it in temperatures of 60° to 75°F.

The wet-cell-storage battery is able to deliver a large amount of power for a short time. After starting, the engine generator or alternator system provides the electrical power for operation of the electrical system and re-charging the battery for the next start.

A lead-acid battery consists of lead plates (positive lead peroxide and negative spongy lead) surrounded by an electrolyte consisting of a mixture of sulfuric acid and water. The sulfuric acid breaks up into positive and negative ions by chemical action when the battery delivers current.

This chemical action will continue until the lead plates are coated with lead sulfate; then no further chemical action can occur and the battery is discharged.

Charging the battery results when current passes in a reverse direction causing the ions to move back into solution, where they combine, re-forming sulfuric acid.

The capacity of the storage battery is measured in ampere-hours. A one-ampere-hour battery can produce a current flow of one amp for one hour before becoming exhausted. The battery will produce any combination of current flow and time that equals the product of the ampere-hour rating.

For example, a 20-ampere-hour battery can deliver 20 amps for one hour or 40 amps for one-half hour, and so on. If a particular airplane requires 400 amps for starting, the battery could turn the starter for three minutes (20 divided by 400 times 60). With cold weather, sluggish oil could easily double this draw. Also, at 32°F the chemical action takes place at a much slower rate that reduces the battery to near one-half its normal capacity.

The battery's total capacity (ampere-hour measurement) depends on the amount of lead plate material it contains. Each time the battery discharges, even a small amount, recharging causes deterioration of the plates. That is, some of the material falls off. This is a normal process, and when insufficient plate material remains, the battery is worn out.

If a battery is not kept fully charged, sulfating (hard-lead-sulfate crystals) occurs, that cannot be easily changed back to active material. Also, a fully charged battery will not freeze.

### Nickel-Cadmium Battery

The nickel-cadmium battery uses a sintered plate, that is, powdered metal is formed into plates. The active materials are nickel hydroxide for the positive plate and cadmium oxide with some iron oxide for the negative plate. The electrolyte is pure potassium hydroxide (KOH) in distilled water. The advantages of nickel-cadmium batteries are:

- *Long life (5-times greater than lead-acid batteries).*
- *No need for frequent addition of water.*
- *Will not freeze, even at -60°F.*
- *No deterioration if left in a discharged state.*

### Dry-cell Battery

Dry-cell batteries are used in the airplane for flashlights, emergency lights, electronic calculators, portable radios, clocks, emergency locator transmitters, and other such items. Regular carbon-dry-cell batteries can lose power rapidly; however, alkaline-dry cells resist loss of power due to freezing temperatures. For this reason, use alkaline cells whenever possible.

If a battery-powered device is used infrequently, reverse one of the batteries to prevent accidental discharge.

## WHEEL COVERS AND WHEEL WELLS

At near-freezing temperatures, mud and slush may be splashed into the wheel covers of fixed-gear airplanes or thrown into the wheel wells of retractable-landing gear airplanes during taxi and takeoff. It then freezes during flight, creating landing gear operation problems.

During cold weather, remove the wheel pants from fixed-gear aircraft to prevent freezing the wheels or brakes.

For retractable-gear airplanes, after takeoff, let the wheels spin to sling off excess moisture and then recycle the gear as a preventive procedure.

## ENGINE PREHEAT

It is possible to start an airplane engine without preheating, even in temperatures as low as -25°F, because of the advent of multi-viscosity oils. This will decrease the life of the engine. Generally, +20°F or lower is considered "preheat" temperature. Below the pour point, oil quickly turns to a molasses-type liquid and any colder temperature results in the oil rapidly approaching a solid.

Besides the change in oil viscosity, batteries lose a high percentage of their effectiveness and instruments stick or freeze.

When you have determined to preheat the engine, adhere to the following:

- *Do not use a heater if it is not in good operating condition.*

- *Do not refuel the heater or the airplane while the engine heater is operating.*
- *Keep a fire extinguisher nearby.*
- *Do not direct heat directly on combustible parts or engine accessories.*

In a sheltered area, the engine may be kept warm with the use of a mechanic's drop cord with a 60- to 100-watt bulb installed. Place this light near the bottom of the engine, inside the cowling. Use an insulated engine blanket (old quilt) when parking the airplane outside.

It is possible to direct the hot air blast from an engine heater at the cylinders and when the engine appears heated and ready to go, the oil is still in a near-frozen state. It is best to direct the hot air blast at the bottom of the oil pan.

When the engine is shut down after a flight, moisture condenses on the inside and outside of the case. Preheating causes these tiny, frozen droplets to melt and disappear. If the preheating continues until the moisture disappears entirely on the outside of the engine and the propeller turns easily, the preheating process is complete.

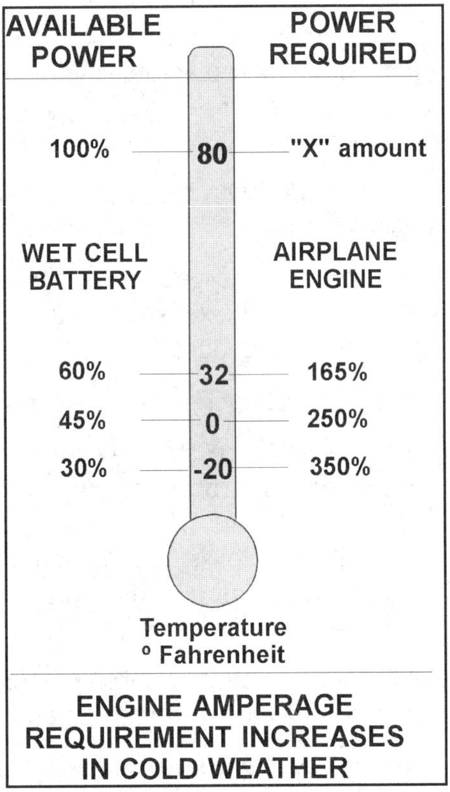


*Figure 5.* Engine battery amperage required vs. available for starting.

## COLD-ENGINE STARTS

Above 20°F the engine may be started without preheat, but the oil may be partially congealed. It is recommended the propeller be turned over by hand at least four times to relieve some of the stress on the starter and battery.

Never operate a starter motor for more than 90 seconds continuously. If the engine does not start, allow a full minute before attempting another start. This allows the starter motor to cool down. After the initial attempt, do not operate the starter for more than 30 seconds continuously.

### Primer

Some pilots have a tendency to over prime during cold starts. This washes down the cylinder walls of oil, since raw fuel is directed into the cylinders without passing through the carburetor and intake manifold. When starting, lack of lubrication causes scoring of the cylinder walls. This in turn causes lack of compression and causes hard starting.

Over priming can also cause induction fires. If a fire occurs, continue operating the starter in an attempt to suck the fire back into the engine where it belongs.

## FROSTED PLUGS

Another problem that plagues engines that have not been preheated is that of "frosted" spark plug electrodes. The pilot tries to start the engine. The engine fires a couple of times and then quits. Any further attempt at starting is futile since there has been sufficient combustion to cause some moisture in the cylinders, but insufficient combustion to heat them up. As a consequence, the water condenses on the spark plug electrodes, freezes to ice, and shorts them out. The only remedy is heat. If a heater is not available, remove the plugs and heat them until all moisture disappears.

For prolonged cold-weather operations, switching to fine-wire spark plugs helps eliminate the frosted-plug problem and improves cold starts.

The author's recommendation is to use fine-wire spark plugs year round. They cost a bit more than the massive-electrode plugs initially, but will save you money on fuel. They also last longer, making them cheaper in the long run.

When the temperature is extremely cold, fuel vaporization becomes a problem. Immediately after starting, apply carburetor heat to help vaporize the fuel. As soon as the engine will run smoothly without carburetor heat, turn the carb heat off.

Engines have a habit of quitting during prolonged idling because insufficient heat is produced to keep the plugs from fouling. Avoid prolonged idling and you'll avoid iced-over or fouled spark plugs.

## OIL PRESSURE INDICATION

You should look for an oil pressure indication within 30 seconds after starting in warm temperatures. There should be an indication within 60 seconds during cold temperatures. If there is no indication, investigate. Turn off the mags, rather than the mixture, to stop the engine faster. Sometimes, even with engine preheating, the oil pressure line to the gauge may contain frozen oil that requires preheating too.

Immediately after the engine starts, operate the engine at an idle speed of 800-1,200 rpm until several minutes of running loosens (warms) the oil to provide lubrication. Revving the engine to as little as 1,500 rpm immediately after starting (even in warm weather) will destroy the engine over a period of time. The pilot who does rev the engine could easily reduce the TBO (time between overhaul) of the engine by half.

Turbine engines can accumulate internal ice overnight and resist rotation when trying to start them. With any indication of locked rotor, unusual noise, or low rpm, discontinue the start. Be aware the rotors can freeze on any cold weather start and be alert enough to stop the start before damaging the engine. If weather forecasts suggest snow, ice or sleet, install engine cowl plugs. This should be done anyway, if the airplane is parked outside, summer or winter.

*Figure 6.* **Note the lack of wheel pants for winter operations.**

# CHAPTER 12

# Regulations

## TAIL WHEEL REGULATIONS

The requirement for a taildragger endorsement contains a grandfather clause. If you have logged pilot-in-command flight time in a taildragger prior to April 15, 1991, the endorsement is not required. For FAA and insurance companies it is wise to have a record of the logbook entry showing this flight time. Otherwise, a one-time logbook endorsement is required.

At the present time FAR Section 61.31(g) states:

> *(i) Additional training required for operating tail wheel airplanes.*
>
> *(1) Except as provided in paragraph (i)(2) of this section, no person may act as pilot in command of a tail wheel airplane unless that person has received and logged flight training from an authorized instructor in a tail wheel airplane and received an endorsement in the person's logbook from an authorized instructor who found the person proficient in the operation of a tail wheel airplane. The flight training must include at least the following maneuvers and procedures:*
>
> *(i) Normal and crosswind takeoffs and landings;*
>
> *(ii) Wheel landings (unless the manufacturer has recommended against such landings); and*
>
> *(iii) Go-around procedures.*
>
> *(2) The training and endorsement required by paragraph (i)(1) of this section is not required if*

> *the person logged pilot-in-command time in a tail wheel airplane before April 15, 1991.*

No person may act as PIC of a tail wheel airplane unless that pilot has received flight instruction from an authorized flight instructor who has found the pilot competent to operate a tail wheel airplane and has made a one-time endorsement so stating in the pilot's logbook.

The endorsement must certify that the pilot is competent in normal and crosswind takeoffs and landings, wheel landings (unless the manufacturer has recommended against wheel landings), and go-around procedures.

The endorsement is not required if a pilot has logged flight time as PIC of tail wheel airplanes before March 15, 1991.

## Recommendation

In addition to the regulatory requirements of Part 61, the FAA recommends that pilots obtain a thorough checkout and transition training for each make and model of tail wheel airplane to be flown, due to significant differences in operating characteristics of individual tail wheel airplanes.

For example, many older types of tail wheel airplanes have pronounced or unusual stall and spin characteristics that differ greatly from those of more recently certificated tail wheel airplanes. In addition, many older airplanes may lack the comprehensive operating data and information typically found in pilot operating handbooks for comparable newer airplanes. Also, systems taken for granted in newer-model airplanes may not exist in older aircraft, requiring a pilot to be familiar with unusual or seldom used procedures.

For example, the absence of electrical systems on many older aircraft compels the pilot to be familiar with hand-propping procedures. The absence of attitude-gyroscopic and heading-gyroscopic instruments requires the pilot to depend more heavily on visual and other cues for basic aircraft control. Finally, the lack of radio equipment in many tail wheel airplanes obligates the pilot to be current in navigation by pilotage and no-radio-traffic-pattern procedures.

Additional factors may affect the instructional environment in tail wheel airplanes equipped with tandem seating. These factors may include reduced visibility from the rear seat, difficulty in communicating with the student due to seating position and higher noise levels (assuming an airplane without an intercom), and lack of complete instrumentation or aircraft controls for the pilot in the rear seat.

Before conducting checkouts or other training in tail wheel airplanes, CFIs should carefully review their own qualifications. Some newly certificated CFIs will have had little or no experience in tail wheel aircraft and will need comprehensive checkouts and transition training in tail wheel airplanes before giving instruction in them.

The recommended endorsement for a pilot to act as PIC in a tail wheel airplane (FAR Section 61.31(d)(2) consists of the following statement:

> *I have given Mr./Ms. ____________, holder of pilot certificate #____________ flight instruction in normal and crosswind takeoffs and landings, wheel landings (if appropriate), and go-around procedures in a tail wheel airplane and find him/her competent to act as PIC in tail wheel airplanes. (Signed:) Date, signature, CFI #, expiration date.*

## RECENT FLIGHT EXPERIENCE

### FAR 61.57 Recent flight experience: Pilot in command

No person may act as pilot in command of an aircraft carrying passengers, nor of an aircraft certificated for more than one required pilot flight crewmember, unless within the preceding 90 days, he has made three takeoffs and three landings as the sole manipulator of the flight controls in an aircraft of the same cat egory and class and, if a type rating is required, of the same type. *If the aircraft is a tail wheel airplane, the landings must have been made to a full stop in a tail wheel airplane.*

## RESOURCE

The author is not associated with Summit Aviation; however, he has used Summit Aviation's Computerized Aviation Reference

Library (on CD-ROM) for years. It includes information about FARs (14 CFR 1-1199), 49 CFR (Hazmat - Hazardous Materials), more than 400 Advisory Circulars, FAA Handbooks, Aviation Weather Services, Training, Maintenance (including ADs), Administration, Pilot Examiner's Handbook, Law, Legal Interpretations, Preambles and more than 5,000 illustrations.

If you are interested in rapidly finding aviation information:

Summit Aviation
Golden, Colo.
1-800-328-6280
303-425-5994

has the *Computerized Aviation Reference Library*, at a reasonable price.

# 1954 Cessna 170B - Pilot's Operating Checklist

## PREFLIGHT CHECK

Legal . . . . . . . . . . . . . . . . . . . ARROW
Ignition Switch (MAGS) . . . . . . . . . . OFF
Control Lock . . . . . . . . . . . . . . REMOVE
Fuel Selector/Quantity . . . . . . . . CHECK
Flaps . . . . . . . . . . . . . . . . . . . EXTEND
External Preflight . . . . . . . . COMPLETE

## BEFORE STARTING ENGINE

Passenger Briefing . . . . . COMPLETE
Seat & Lock . . . . ADJUST & INSTALL
Seat Belt/Shoulder Harness . . . . ON
Brakes . . . . . . . . . . . . . . TEST & SET
Radios & Electrical . . . . . . . . . . . OFF

## STARTING ENGINE

Mixture . . . . . . . . . . . . . . . . . . . . RICH
Propeller . . . . . . . . . . . . . . . HIGH RPM
Carburetor Heat . . . . . . . . . . . . . COLD
Throttle . . . . . . . . . . . . . . . . . CRACKED
Prime . . . . . . . . . . . . . . . AS REQUIRED
Master & Beacon . . . . . . . . . . . . . . ON
Propeller Area . . . . . . . . . . . . . CLEAR
Ignition Switch . . . . . . . . . . . . . START
Oil Pressure . . . . . . . . . . . . . . . CHECK
Avionics & Electrical . . . . . . . . ON

## TAXI CHECK

Wing Flaps . . . . . . . . . . AS REQUIRED
Controls . . . . . . . . . FREE & CORRECT
Trim . . . . . . . . . . . SET FOR TAKEOFF
Instruments & Radios . . . . . . . . . . SET
Amps & Suction Gauge . . . . . . CHECK
Circuit Breakers . . . . . . . . . . . . CHECK

## BEFORE TAKEOFF

Throttle . . . . . . . . SET 1,800 RPM
Mixture . . . . FOR DENSITY ALTITUDE
Magnetos . . . . . . . . . . . . . . CHECK
Carburetor Heat . . . . . . . . CHECK
Propeller . . . . . . . . . . . . . . CYCLE
Fuel Pump . . . . . . . . . . . . . CHECK

## NORMAL TAKEOFF

Strobe Lights . . . . . . . . . . . ON
Transponder . . . . . . . . . . . ON
Landing Light . . . . . . AS REQUIRED
Climb . . . . . . . . . 85 to 78 MPH

## SHORT-FIELD TAKEOFF

Strobe Lights . . . . . . . . . . . ON
Transponder . . . . . . . . . . . ON
Liftoff & Climb . . . . 62 MPH (flaps up)
Liftoff & Climb . . . . 58 MPH (flaps 20°)

## AFTER TAKEOFF

Fuel Pump . . . . . . . . OFF (if used)
Landing Light . . . . . . . . . . . . OFF

## EN ROUTE CLIMB

Airspeed . . . . . . . 80-100 MPH IAS
Throttle . . . . . . . . . . . FULL OPEN
Propeller . . . . 2,300 to 2,400 RPM
Mixture . . . . . DENSITY ALTITUDE

## CRUISE

Power . . . . . . . . . ADJUST 65-75%
Mixture . . . . . . . . . . . . . . ADJUST
Heading Indicator . . . . . . . CHECK

## DESCENT

Mixture . . . . . . . . . . . . . . . ADJUST
Carburetor Heat . . . . AS REQUIRED
Landing Light . . . . . . AS REQUIRED

## BEFORE LANDING

Seat Rail . . . . . . LOCK INSTALLED
Belts & Harness . . ADJUST & LOCK
Fuel Selector . . . . . . . . . . . BOTH
Landing Light . . . . . . . . . . . . . ON
Mixture . . FOR DENSITY ALTITUDE
Carburetor Heat . . . AS REQUIRED
Flaps . . . . . . . . . . . AS REQUIRED
Airspeed . . . . . . 70 MPH MAXIMUM
Propeller . . . . . ADJUST HIGH RPM

## AFTER LANDING

Wing Flaps . . . . . . . . . . UP
Carburetor Heat . . . . . . . . . . COLD
Landing Light . . . . . . . . . . OFF
Strobe Lights . . . . . . . . . . OFF
Transponder . . . . . . . . . . STANDBY
Trim . . . . . . . . . . SET FOR TAKEOFF

## STOPPING ENGINE

Avionics & Electrical . . . . . . . . . . OFF
Throttle . . . . . . . . . . IDLE
Mixture . . . . . . . . . . IDLE CUT-OFF
Master . . . . . . . . . . OFF
Magnetos . . . . . . . . . . OFF
Brakes . . . . . . . . . . AS REQUIRED
Controls . . . . . . . . . . LOCKED
Hobbs Meter . . . . . . . . . . RECORD
Pitot Cover . . . . . . . . . . INSTALL
Tie-down . . . . . . . . . . COMPLETE

## AIRSPEEDS FOR SAFE OPERATION

Vso (stall speed, flaps) . . . . 55 mph
Vs1 (stall speed, no flaps) . . 59 mph
Vx (best angle-of-climb) . . . . 62 mph
Vy (best rate-of-climb) . . . . . 85 mph
Va (maneuvering speed) . . . 115 mph
Vfe (max flaps extended) . . . 100 mph
Vno (max structural cruise) . . 140 mph
Vne (never exceed) . . . . . . 160 mph

## ELECTRICAL FIRE

Master . . . . . . . . . . OFF
Electrical & Radios . . . . . . . . . . OFF
Air Vents . . . . . . . . . . CLOSED
Fire Extinguisher . . . . AS REQUIRED
Circuit Breakers . . . . . . . . . . CHECK
Master . . . . . . . . . . ON, IF REQUIRED

If a circuit breaker is reset and "pops" out, do not reset again.

## ENGINE FAILURE

Establish Glide . . . . . . . . . .70-75 MPH
Cockpit Check . . . . . . . . . . PERFORM
Fuel Selector . . . . . CHANGE TANKS
Mixture . . . . . . . . . . RICH
Carburetor Heat . . . . . . . . . . ON
Fuel Pump . . . . . . . . . . ON
Primer . . . . . . . . . . LOCKED
Restart . . . . . . . . . . ATTEMPT
Radio . . . . . POSITION & SITUATION

## FORCED LANDING

Airspeed . . . . . . . . . . 70-75 MPH
Mixture . . . . . . . . . . IDLE CUT-OFF
Fuel Selector . . . . . . . . . . OFF
Magnetos . . . . . . . . . . OFF
Flaps . . . . . . . . . . FULL (for landing)
Seat Belt & Harness . . . . . TIGHTEN
Doors . . . . . . . . . . UNLATCHED
Radio . . . . POSITION & INTENTIONS

## ENGINE FIRE

Mixture . . . . . . . . . . IDLE CUT-OFF
Fuel Selector . . . . . . . . . . OFF
Master . . . . . . . . . . OFF
Air Vents . . . . . . . . . . CLOSED
Airspeed . . . . . . . . . . 140 MPH*

* to blow out the fire

## WEIGHT & BALANCE

Gross weight . . . . . . . . . . 2,200 lbs
Empty weight . . . . . . . . 1,522.75 lbs
Empty index . . . . . . . . . . 56.6"

## MISCELLANEOUS

Fuel . . . . . . . . . . 37 gals usable
Fuel burn . . . . . . . . . . 8.7 gph
True airspeed . . . . . . . . . . 112 knots

# V SPEEDS

Airspeed is defined as the velocity of the airplane relative to the air mass through which it is moving. Groundspeed is different from airspeed. When an air mass is moving, the combined effect of airspeed and air mass speed determine the ground speed.

On page 8-11 the following was given as a "Remember" item:

*A constant attitude and a constant power setting results in a constant performance. This is true regardless of the phase of flight such as climbing, cruising or gliding.*

What this means is that pitch and power determine airspeed. The FAA has defined airspeeds with abbreviations representing velocity; hence the term "V" speed.

$V_A$ means design maneuvering speed.

$V_B$ means design speed for maximum gust intensity.

$V_C$ means design cruising speed.

$V_D$ means design diving speed.

$V_{DF/MDF}$ means demonstrated flight diving speed.

$V_{EF}$ means the speed at which the critical engine is assumed to fail during takeoff.

$V_F$ means design flap speed.

$V_{FC/MFC}$ means maximum speed for stability characteristics.

$V_{FE}$ means maximum flap extended speed.

$V_{FTO}$ means final takeoff speed.

$V_H$ means maximum speed in level flight with maximum continuous power.

$V_{LE}$ means maximum landing gear extended speed.

$V_{LO}$ means maximum landing gear operating speed.

$V_{LOF}$ means liftoff speed.

$V_{MC}$ means minimum control speed with the critical engine inoperative.

$V_{MO/MMO}$ means maximum operating limit speed.

$V_{MU}$ means minimum unstick speed.

$V_{NE}$ means never exceed speed.

$V_{NO}$ means maximum structural cruising speed.

$V_R$ means rotation speed.

$V_{REF}$ means reference landing speed.

$V_S$ means the stalling speed or the minimum steady flight speed at which the airplane is controllable.

$V_{SO}$ means the stalling speed or the minimum steady flight speed in the landing configuration.

$V_{S1}$ means the stalling speed or the minimum steady flight speed obtained in a specific configuration.

$V_{SR}$ means reference stall speed.

$V_{SR0}$ means reference stall speed in the landing configuration.

$V_{SR1}$ means reference stall speed in a specific configuration.

$V_{SW}$ means speed at which onset of natural or artificial stall warning occurs.

$V_{TOSS}$ means takeoff safety speed for Category A rotorcraft.

$V_X$ means speed for best angle of climb.

$V_Y$ means speed for best rate of climb.

$V_1$ means the maximum speed in the takeoff at which the pilot must take the first action (e.g., apply brakes, reduce thrust, deploy speed brakes) to stop the airplane within the accelerate-stop distance. V1 also means the minimum speed in the takeoff, following a failure of the critical engine at VEF, at which the pilot can continue the takeoff and achieve the required height above the takeoff surface within the takeoff distance.

V2 means takeoff safety speed.

$V_{2min}$ means minimum takeoff safety speed.

## Flight Planning

### Rule of Thumb — Modify Climb Time

*Add one minute for each 1,000 feet of climb altitude.*

*Compute the altitude difference between the departure elevation and the cruise altitude. Use the computed ground speed for time and distance calculations along the entire distance. Add one minute per 1,000 feet of climb to the en route time for the first leg to compensate for the reduced true airspeed during the climb.*

*If your airplane has greater performance than most general aviation planes, you may find one minute too much. If so, try adding one minute for each 2,000 feet of climb. If this doesn't work out, modify it to suit your airplane.*

*Use computed ground speed to calculate flight time. Add one minute per 1,000 feet of climb to the en route time for the first leg.*

### Rule of Thumb — Cruise Altitude

*Do not climb more than 10 minutes for each hour of en route flight time. If the terrain dictates a higher altitude than that specified by this rule, you will have to sacrifice efficiency.*

*Depending upon the performance of your airplane, it is usually inefficient to climb more than 10 minutes per hour of estimated time en route for the entire trip.*

*If the rate of climb in your airplane is 500 feet per minute, you shouldn't climb more than 5,000 feet above ground level for each one hour of flight time.*

### Rule of Thumb — Convert Celsius to Fahrenheit

*Double the Celsius temperature and add 30°. This rule is only an approximation. For an exact conver-*

*sion, double the temperature in Celsius, subtract 10% and add 32. Example: 15°C = ? °F.*

*15 + 15 = 30 –10% = 27 + 32 = 59°F*

### Rule of Thumb – Humidity

*There is no rule of thumb for determining the effect of humidity on aircraft performance. If high humidity exists, add 10 percent to the computed takeoff distance. Anticipate a reduced rate of climb.*

### Rule Of Thumb – Centrifugal Force During A Swerve

*Centrifugal force increases as the square of the speed at which it starts.*

## Performance

### Rule of Thumb – Increased Takeoff Distance

*For each 1,000 feet above sea level, the takeoff run will increase approximately 12 percent.*

### Rule of Thumb – Change in Weight/Stall Speed

*A two-percent change in weight causes a one-percent change in stall speed.*

### Rule of Thumb – Stall Speed Increase

*The stall speed increases as the square root of the wing load factor.*

*For example, in a 60-degree bank the load factor is 2 Gs. The square root of two is 1.41, resulting in a 41 percent increase in stall speed.*

### Rule of Thumb – Ground Effect

*When trying to accelerate after a soft-field or short-field takeoff, maintain the wing as close to the ground as possible. During the acceleration to VX or VY, fly no higher than one-half the wing span above the ground.*

### Rule of Thumb — Determine Density Altitude

*For each 10°F above (or below) standard temperature at an airport's elevation, add (or subtract) 600 feet to (from) the field's elevation.*

## GROSS WEIGHT

*The takeoff distance varies with gross weight. A 10-percent increase in takeoff gross weight (while not exceeding the maximum allowable gross) will cause a:*

- *5-percent increase in the speed necessary for takeoff.*
- *9-percent decrease in acceleration to takeoff speed.*
- *21-percent increase in the takeoff distance.*

### RULE OF THUMB — Takeoff Distance vs. Gross Weight

*The takeoff distance varies with the square of the gross weight. Whenever the gross weight of an airplane is increased, the performance is decreased. The takeoff distance varies with the square of the gross weight. This makes sense. If the gross weight of the airplane is increased, it is logical that the takeoff speed must be increased to compensate for the additional weight.*

## Arrival

### Rule of Thumb — Computing Descent

*3 x altitude (in hundreds of feet) equals distance out to begin descent.*

*One-half the ground speed (in knots) times 10 = rate of descent for 3-degree glide path.*

## Landing

### Rule of Thumb — Landing Flight Path Adjustments

*Maintain the airspeed with elevator and altitude with throttle.*

### Rule of Thumb – Increased Landing Speed

*The landing speed (true airspeed) will increase about 2 percent for each 1,000 feet above sea level when using the same indicated airspeed for approach as at a sea level airport.*

### Rule of Thumb – Approach Indicated Airspeed

*Regardless of altitude, ALWAYS fly your approach for landing at the normal sea-level approach indicated airspeed for the airplane.*

### Rule of Thumb – Approach Speed and Gross Weight

*Reduce the calibrated approach airspeed for the maximum landing weight listed by one half of the percentage of the weight decrease.*

*The percentage of weight decrease is the ratio of the maximum allowable gross weight divided by the actual operating weight. In the case of the Cessna 182, the maximum allowable gross weight is 2,950 pounds. If the airplane is being operated at 2,500 pounds, the weight decrease is 450 pounds. The percentage of weight decrease is 450 divided by 2,950 for 0.1525, or a 15.25 percent decrease in gross weight. The maximum allowable gross weight calculated approach speed, 70 mph, should be reduced by half the percentage of weight decrease (7.6 percent). The optimum approach speed at a reduced weight of 2,500 pounds is 65 mph (70-(70*1.076) = 65 mph).*

### Rule of Thumb – High-Speed Descent

*Before making a high-speed descent, check for pilot reports concerning the ride during descent. If not available, study the winds aloft forecast for a change in wind direction occurring throughout the altitudes of the operation. Check also for wind velocity changes. Usually a change of six knots per 1,000 feet of altitude change is sufficient to produce turbulence.*

# Index

## A



## B



## C

## D

## E



## F

## G



## H



## I

## J



## K



## L



## M

## N



## O



## P

## S

## T